Keys to the Kingdom of Alchemy

Keys to the Kingdom of Alchemy

Unlocking the Secrets of Basil Valentine's Stone

J. Erik LaPort

Quintessence

KEYS TO THE KINGDOM OF ALCHEMY
by J. Erik LaPort
Copyright © 2015 Quintessence

Paperback Color Edition ISBN: 978-0-9906198-4-0

All Rights Reserved. No part of this publication may be commercially reproduced, stored in a retrieval system, or transmitted, in any form or in any means – by electronic, mechanical, photocopying, recording or otherwise – without prior written permission from the authors apart from fair-use specified below.

The publisher permits and encourages Fair Use of the material contained in this book for the noncommercial purposes of teaching, research, scholarship, non-profit education, factual or non-fiction publications, criticism or commentary.

First English edition published 2015 by Quintessence

Publication Assistance and Digital Printing in the United States, Australia and/or the UK by Lightning Source®

Initial Edit by Victor Hörnfeldt

Final Edit by Moshe Daniel Block ND, HMC
www.david-house-productions.com

Cover design by Rachel Turton
www.graphicdesignkohtao.com

Paperback and Hardcover Editions Distributed by Ingram

Digital Edition distributed by Third-Order Alchemy
www.third-order-alchemy.com

Images Attribution: Chemical Heritage Foundation

The Fell Types are digitally reproduced by Igino Marini
www.iginomarini.com

DEDICATION

To every alchemy researcher who has reached the 4th *Key* and hit that impenetrable wall, this book is for you.

CONTENTS

Acknowledgments i
Foreword iii
Author's Preface v
Editor's Note ix
Unlocking the Gates xix
Key I   – Grey Wolf and Lion King 1
Key II   – Two Mercuries 7
Key III   – Cœlestial Sulphur 15
Key IV   – Salt of Ashes 27
Key V   – Spirit of Mercury 37
Key VI   – Chemical-Wedding 49
Key VII   – Judged by Fire 57
Key VIII – Putrefaction of Adam 65
Key IX   – Planetary Colors 73
Key X   – Water, Ashes and Sand 83
Key XI   – Noble Offspring 93
Key XII   – Fermented Tincture and Fixt Gold 101
Appendix – Cover-terms used by Valentine 107
Author's Postface 109

ACKNOWLEDGMENTS

This book would not have been possible without the understanding and patience of my loving wife Sunisa Kaewyindee (Paw), and Victor Hörnfeldt, who lent patronage and steadfast support throughout all stages of research and writing of this book.

As the source text for this interpretation, I have chosen the 1670 and 1671 English edition compiled and printed by S.G. and B.G. for Edward Brewster, and available in the public domain online from the Internet Archive and the Open Library.

Special thanks to my content editor, Moshe Daniel Block, whose keen eye and razor-sharp penetrating insights greatly improved this work as a whole. Gratitude also to Steve Kalec, an adept alchemist with the patience and capabilities of the alchemists of yore, for agreeing to write the Foreword – his kind encouragement and support during a moment of doubt bolstered my resolve to remain true to the original interpretation herein.

The images chosen for commentary are the woodblock images engraved by Matthäus Merian for publisher Johann Theodor de Bry, published in the collection *Musaeum hermeticum*, 1678 and placed in the public domain by the Chemical Heritage Foundation.

Thanks to Rachel Turton for her positive attitude and artistic input. It is always a pleasure working with her and we feel lucky that someone of her energy and talent lives on our tiny island in the middle of nowhere.

Finally, to my close friend Roger Gabrielsson for his patience in helping me to understand the finer points of chemistry as they pertain to my interest in al/chemical interpretation, eternal thanks indeed.

FOREWORD

For many, the writings of the ancient Alchemists, often times obscured with their many philosophical and mythological allegories and metaphors, are indeed a challenging study and difficult practical undertaking. In fact, the alchemists do tell us that they write for their own kind and not for the common man, and they warn us to be careful of the literal interpretations and understanding. They wrote and described the many steps in their works, yet have kept their words and instructions most cryptic. However, it is also written that one book opens another book, that the words of one master reveal the words of another, that one inspiration begets another, and one flame will ignite another. *Keys to the Kingdom of Alchemy* is one such valuable book. Therein are found the author's amazing insights that reveal the words of Basil Valentine. It is a book that inspires and illuminates the quest for the so much sought after Philosopher's Stone of the ancients.

Basil Valentine certainly was very advanced in the chemical knowledge of his time. It is most clearly shown by the author that Valentine's creative thinking, inventive and ingenious methods of work and experimentation, lead him to discover new alchemical products and processes that surpassed the more ancient alchemical traditions. Valentine, being a wise philosopher, understood the alchemical allegories and metaphors, which enabled him to work as a great researcher, scientist and chemist. He became one of the great alchemists whose writings all aspiring alchemists truly seek to understand, and whose works they seek to master.

Keys to the Kingdom of Alchemy is a major work and will be found to be most valuable to all serious researchers in the study of *The Twelve Keys of Basil Valentine*. It is quite evident that Erik LaPort has completed very extensive research, and that he is an authority on ancient laboratory technologies and processes. Having himself a keen understanding of the philosophical and mythological allegories, he most convincingly reveals to us the sometimes cryptic words and instructions concerning the Archetypal Alchemical Stone of Basil Valentine as is set down for us in his *Twelve Keys*.

To the alchemists, their matters were considered philosophical matters as they were philosophically prepared and so they were not their common kind. One will truly appreciate how this book so instructively reveals the various preparations of these matters to produce other alchemical matters which lead us to the final preparation of the Alchemical Archetypal Stone. Key by key, we are brought to light through the author's expertise in ancient laboratory techniques and procedures. Being a passionate researcher into ancient technologies and an avid experimenter, Erik LaPort divulges for us the laboratory methods and alchemical secrets being hidden in this ancient and most important alchemical work by Basil Valentine.

The Archetypal Stone of Basil Valentine as revealed in his *Twelve Keys* is the *Summum Bonum* of alchemical works of the alchemists. Revelations and references to Valentine's work on *Phalaja* and the *Fire Stone* are true gems to cherish. Whether one is a researcher or a student, an aspiring or practicing alchemist, he will find *Keys to the Kingdom of Alchemy* an inspirational and motivating book full of revelation and instruction, shedding illuminating light on the great works of Basil Valentine.

Steve Kalec

PREFACE

Basil Valentine's *The XII Keys* is the "how-to" section of a book entitled *On the Great Stone of the Ancient Philosophers*. It is an encrypted alchemical text that preserves the exact recipe for confecting the Philosophers' Stone of archetypal alchemy ... but only if each Key can be decoded. Perhaps others have already succeeded in unlocking its mysteries, but if this were the case, such work has never been made public. Many researchers throughout history have succeeded in opening the first three gates, but they were all halted at the fourth.

After spending years researching the archetypal recipes and chemical identities for the White Stone (Hermes) and the Tincture (Ios; al-Iksir) of Alexandrian alchemical tradition, I became increasingly interested in discovering whether these products were faithfully reproduced by European alchemists. Valentine was consistent throughout primary and supporting texts that he was addressing the *Great Stone of the Ancient Philosophers*. It was this specific claim that compelled me to attempt to either find supporting evidence to either verify or dismiss such a bold claim, and state the grounds for doing so.

High ideals aside and with abject honesty, growing curiosity became the primary motivating factor for interpreting *The XII Keys*. I was unprepared for what I would discover – if my interpretations were correct, Valentine not only reproduced the *Tincture of the Philosophers*, he actually presented exact ratios and optimized the recipe for its confection. I felt that a new interpretation might be of tremendous interest to earnest researchers studying archetypal alchemy. I quickly realized that although the author of *The XII Keys* was addressing a precise chemical reaction, he viewed confecting the stone as a complex material representation of interwoven ideologies.

Keys to the Kingdom of Alchemy primarily addresses operative alchemy. Apart from operative aspects however, a tremendous amount of rich allegorical storytelling and masterful conceptual blending reveals the author of *The XII Keys* to have been a highly educated polymath fluent in not only alchemical symbolism but also Christian mysticism, Greco-Roman mythology and Perennial Philosophy. Such a complete package in a single author is certain to be of interest to researchers attracted to the speculative aspects of alchemy.

An understanding of the archetypal Alexandrian model (chrysopœia) serves as a type of master key that unlocks alchemical secrets, this being known to many European alchemists who pursued the Stone. My own understanding of the chemical processes used in early alchemical traditions, for which I am indebted to my good friend Dr. Roger Gabrielsson, provided the theoretical framework prerequisite to attempt a new interpretation of *The XII Keys*.

Progressing through the work led to encouragement and inspiration upon discovering, if the interpretation herein is correct, that the original author of Basil Valentine's *The XII Keys* was true to his claim of delivering the recipe for the *Great Stone of the Ancient Philosophers*, and not a new European variant. The work unfolded into a coherent and linked series of chemical preparations and processes beautifully interwoven with complex allegory and symbolism. Journeying through the recipe while immersed in the mindset of the master alchemist who authored *The XII Keys*, whomever Basil Valentine actually was, has been an absolute joy.

This interpretation was completed in a relatively short time, which was admittedly made much easier due to our ongoing alchemical research and reproduction beginning many years prior. I sincerely hope that this new interpretation becomes a welcome addition to the enduring body of research into the history of alchemy and chemistry. Admittedly, no single treatment is ever perfect, this work is certain to contain errors, which academic scholars and independent researchers will ultimately correct. To them I would like to express my deepest gratitude in advance for furthering the subject.

Valentine states clearly in a number of instances that one Key proceeds from another and therefore this book is best read chapter-by-chapter in a pedagogical order. I include quotations from supporting alchemical texts attributed to Basil Valentine, yet some of these may have been forgeries or written by a collective Pseudo-Valentine, a fact that raises questions only as regards those particular statements. By focusing exclusively on the instructions drawn from *The XII Keys*, the chemistry strongly supports Valentine's claim that he was presenting the recipe for the archetypal *Great Stone of the Ancient Philosophers*.

J. Erik LaPort

EDITOR'S NOTE

Editing Erik LaPort's book *Keys to the Kingdom of Alchemy* has been a great joy and quite an amazing experience for me. I could describe it as quite unconventional for several reasons. First reason being, I am not really an editor, although the four books I have written and seen edited have helped me at least learn the basics of the role of the editor. Erik was gracious to share a review copy of his book and as I read, I began finding errors in it and made suggestions. Erik liked some of my suggestions and we continued from there. He, himself, is also very unconventional as a writer and as a researcher into alchemy (he says he's not an alchemist). If he were a conventional alchemist, we wouldn't be receiving such unveiled, revelatory insights into Basil Valentine's work.

Being an alchemist myself, for many years I've been hitting my head against the ubiquitous agreement of secrecy found in all alchemical texts. As I read this book, my eyes grew wider and wider in amazement at the transparency with which Erik was unfolding the secrets in the Keys. That is most unconventional for an author on alchemy. I've also read his book *Cracking the Philosopher's Stone* and was equally amazed at how he has brought together research across millennia of alchemy's history, demonstrating a common thread that has run through various processes leading to the archetypal Philosopher's Stone. If one delves into the research of this book and embraces Erik's conclusions, it not only cracks open Basil Valentine's highly vaulted works (beyond strictly his *Twelve Keys*), but also begins elucidating many other alchemical manuscripts that adhered to a mostly universal approach to the archetypal Philosopher's Stone.

Notice how I am writing, "Philosopher's Stone." That's the traditional way. Erik writes it in this work and in his other books as, "Philosophers' Stone". Note also how I put the punctuation within the quotation, which is a modern editorial standard in this day and age. Notice where I put it for Erik's version of "Philosophers' Stone". Outside the quotations, which is how he likes to do it. When I pointed out that he was not doing it according to commonly practiced standards, his reply made me laugh, as it contained research into the historical precedents for using the quotes outside the quotation marks vs inside. This is an insight into his nature and also parallels his passionate desire to get to the bottom of things. He replied by quoting his grammar source, Mignon Fogarty:

> I use the British method: "Compositors—people who layout printed material with type—made the original rule that placed periods and commas inside quotation marks to protect the small metal pieces of type from breaking off the end of the sentence. The quotation marks protected the commas and periods. In the early 1900s, it appears that the Fowler brothers (who wrote a famous British style guide called The King's English) began lobbying to make the rules more about logic and less about the mechanics of typesetting. They won the British battle, but Americans didn't adopt the change. That's why we have different styles."

So I recognized that his outside-the-box choices are not based on whimsical fancy, but come from well thought out intentional design. In addition to that, Erik's response to my attempt at correcting his unconventional form of "Philosophers' Stone" was also quite revealing of Erik's character. He said:

> I've always written it in the plural, and throughout Cracking the Philosophers' Stone as well. It has belonged to, and does belong to, more than one Philosopher as far as I'm concerned. F*** convention in this case. I like my way better.

And that's Erik. Thinking of all the many philosophers and not just the one and putting unnecessary convention up for calcination.

As a practicing lab and spiritual alchemist, I've struggled with the question "To reveal or not to reveal?" And I've flip-flopped back and forth. When I first began on the path of alchemy, I was very much in an innocent state of thinking, "It's the day and age to reveal everything!" But getting deeply immersed in the texts of ancient authors sworn to secrecy, I found myself being swayed by their general thinking. Get something repeated enough to you and it's easy to begin believing it. The thing about Erik is, he doesn't believe any of it, because he comes at it from a very unique perspective. As a "non-alchemist" brilliant historical researcher, and a knowledgeable [hobby] chemist, he approaches it from a place of seeking answers to the riddles of the Adepts for the sheer joy and curiosity of it.

He is the most generous author on alchemy I have ever met. His big heart and desire to share what he has learned has truly helped me flop back to my original thinking entering into this gripping and fantastic field of Natural philosophy. The Foreword to this book has been written by another great modern day alchemist, Steve Kalec, who is another extremely generous and forthcoming person with his knowledge of alchemy. Like Erik, Steve also shares quite a lot more than the average person does. And it's all so that others may benefit. I am grateful to have played even a small role alongside such big-hearted gentlemen in a movement of alchemy that I consider to be fulfilling some of the prophecies of old: The Truth will be shouted from the mountain tops. And from Kabbalah's perspective that in this day and age, "The floodgates of knowledge would be opened." And why not? Why not reveal in this day and age? I've found myself recently on forums arguing *against* revealing so openly, regurgitating the same views I had read over and over. And that's all it is. Regurgitation of views that ultimately are not even my own, nor do they necessarily serve any purpose *today*. Erik doesn't regurgitate. He considers, takes things apart, tests them in the lab, and then reports on his findings. Or, he does a lot of research and presents it as a strong possibility for what may have been.

Don't get me wrong. I can understand why there has been secrecy in the past. If you were an alchemist living during the many epochs when alchemy was a crime punishable by death, you'd have kept the fumes of your Athanor to yourself too, or at least published under a pseudonym if you were to share anything. Another element of withholding information that I do understand is so that the student can actively seek and find answers for themselves, which is tremendously rewarding. However, as Cyliani writes in his work *Hermes Unveiled*:

> *If you could only know, as I do, the various hardships that I have undergone in order to reach this goal, you would draw back in fear. Start only if God allows you to meet a man who has already succeeded in making the Stone. One who will lead you by the hand from the beginning to the end. Push away from your mind in horror the idea of devoting yourself to Hermetic Philosophy. Whole secrets are unbelievably far more difficult to find on one's own. If, hoping to have better luck than I did, you reject my advice and are so fortunate as to succeed, never forget those more unfortunate than yourself.*

Now, with interest in alchemy ever increasing, why let the struggling souls like me, without a master, who grope around in the dark without a candle and compass to show the way to the Stone, continue smashing into the walls of secrecy? Erik LaPort, through this book, becomes the man leading you by the hand. Certainly, knowledge of lab work and chemistry is also crucial in arriving at the final Stone. But what was not permitted to be shared in the "boys' club" of the alchemical Adepts of old, is now accessible in Erik's work.

Since I began getting to know Erik through our correspondences, I've been quite impressed by his detachment to the conclusions he arrives at in his own work. For him, it's all about the research and helping things move forward. If someone found his conclusions to be wrong, I honestly believe Erik would be delighted to see the research brought forward if the person could do so. Whatever that streak of jealousy or possessiveness that is attributed to alchemists regarding their knowledge, and that is likely one of the main reasons for keeping their secrets to themselves, Erik seems to be free of it. This shares insight into his nature, since all people in possession of secret knowledge have to battle the demons that would have them keep it to themselves. He didn't even think of referencing his other book, *Cracking the Philosopher's Stone*, something that, as his editor, I had to point out to him, and I suggested where it would be appropriate, for instance in the places where he refers to Alexandrian alchemy and alchemical gold.

Another editorial discussion that surfaced was concerning certain traditional alchemical views on topics such as "The Secret Fire," and "multiplying the Stone." Erik doesn't hold the stance that I do as an alchemist; that there is 'magic' in alchemy that can bring about the transmutation of matter so that atomic structures of metal, for example, are actually fundamentally changed in their composition into other elements. For Erik, it is strictly chemistry. This is not to say that he is not deeply spiritual and deeply connected and devoted to the Divine. He most definitely is and I think it is his self-healing, his deep self-reflection work, and his consumption of the alchemical products he makes that has made him such a generous and big-hearted author.

It is also not to say that he is correct in all his views on alchemy, and he is the first to say so. He just doesn't believe that the finished Stone can transmute lead or mercury into gold, and holds another view on the gold that was created in the process of "projection" (alchemical gold). He discusses this at even greater length in his book *Cracking the Philosopher's Stone*. So you will notice in this book that Erik doesn't address the idea in the main body of the book that multiplying the Stone also involves augmenting its power to transmute. You won't see talk of the *Secret Fire* (he called it a "can of worms" as it opens up a lot of avenues that are difficult to interpret) as it pertains to an energy beyond mechanical agitation or the periodic table of elements that can slowly bring about fundamental changes in the makeup of matter. But you will see plainly revealed, *The Mercury of the Philosophers* – the most well-kept secret of all of alchemy, without which one cannot proceed to the Stone. You will even understand things like Paracelsus' *Blood of the Red Lion* and *Gluten of the Eagle*. So I suggest not even attempting to pigeon hole this book or the author's writing in any sort of categorical way. It is a gift of a most unconventional and generous kind.

One thing I found interesting was working with Erik on interpreting the Hebrew of the wood carving artwork of the *Tenth Key*. It is filled with mistakes. Whoever wrote it, be it Basil Valentine himself, or the artist that was entrusted to copy his sketch, evidently did not know Hebrew. For example, in the top left of the image:

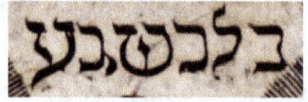

The phrase, (in the heart of Seven) בלב שבע, is spelled incorrectly, with a *gimmel* instead of a *bet*, ב, in the word "seven" שבע. Even the word, *ba'lev*, בלב, has a *kaf* instead of a second *bet*, ב, and the first *bet* looks questionable as well. Also, the word in the top right:

Erik interpreted it to mean "Lunar" ירחי. This is brilliant, because knowing Hebrew, the *yud*, י, should be significantly smaller than the *resh*, ר. And even so, the final *yud* looks almost identical to the *resh*, making it a very hard to understand word. However, it is easy to accept since *Yerechi*, meaning *lunar*, sits right above the symbol of the moon. The most difficult interpretation we had was of the central word:

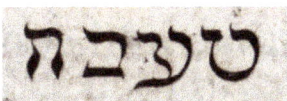

There is really nothing that resembles this in Hebrew that has any real significance to the *Tenth Key*. I thought perhaps the artist made a mistake replacing the *bet*, ב, in this word instead of a *nun*, נ, leaving the word טענה. However, the meaning of the word *Ta'ana* "claim, assertion, argument, reason, plea," doesn't really fit within the context of the *Tenth Key*. So I finally interpreted it to be quite a misspelled version of the word *"Teva"* טבע, meaning "nature, substance, element," which Erik liked and felt it had significance for the Key. Ultimately, the Hebrew on this Key adds very little importance to the work and does not a thing toward the overall interpretation of the Key, so at the least, we were not grappling with anything vital to the overall meaning of the work.

My role here as editor was to work with Erik's creative original style that is outside of conventional, traditional, and regurgitative rules and also to ensure his text was clear, consistent and fulfilling an easy to understand message to the reader. It has been a great honour for me to contribute in this way to the book, and a privilege, as it was really I that was like a kid in a candy store feasting on all the delicious goods, while from time to time pointing out a misplaced candy here or a jar that was placed on the wrong shelf there when I spotted it.

Moshe Daniel Block, ND

... the ancients knew many good things,
for mine own part I must confess,
that I borrowed the foundation of my knowledge from them,
which made me to lay it to heart,
and am thereby moved to leave for others also a corner-stone,
that truth may further be confirmed,
and the grounds of it made easier, clearer, plainer,
and more manifest by a further knowledge of my writing.

Basil Valentine

Unlocking the Gates

... in metals and minerals, of them we may learn, as the twelve Sybils prophesied of the bright, true, and only Son of Righteousness and Truth, in which do rest after the twelve ports and gates of Heaven ...

Basil Valentine

Sometime during 1599, an alchemical text appeared entitled *A Short Summary Tract: Of the Great Stone of the Ancient Philosophers*, which presumably revealed twelve crucial Keys to unlock the secrets of confecting the Philosophers' Stone, its multiplication and application towards alchemical gold making. The work was attributed to Basil Valentine, a name that remains as much an enigma today as it was at the time of publication. There has been much speculation as to the identity of Basil Valentine and the period in which he operated. He was alleged to be a Benedictine monk residing at the Priory of St. Peter at Erfurt, Germany. Accounts of his life, details regarding his monastic life and his homeland along with specific dates occur in writings attributed to him, yet many of these have been determined forgeries. Numerous telltale signs found in the Valentine texts – references to zinc, the plague, the printing press, the French disease (syphilis), tobacco and America to name a few – provide a "written after ... " date that does not support earlier claims. No monk with the name Basilius Valentinus has ever been found in the General Register of the Benedictines in Rome. Yet paradoxically, there once was evidence of his existence. Queen Christina of Sweden allegedly removed a number of writings attributed to Basil Valentine and these have apparently never resurfaced to the public. Anecdotal evidence suggests that St. Peter's Abbey once displayed a picture of Basil Valentine in the Auditorium Philosophicum and that an alchemical laboratory had existed prior to renovations. Dr. D.A. Schein in *Basilius Valentinus and his Tinctures from Antimony* provides a thorough review of the discussion regarding the Valentine corpus author/s' personal identity, which to-date remains unresolved. According to Schein, evidence suggests that the collective works of Basil Valentine were composed during the 16th century. I have used the name Valentine throughout this treatment as a convention to refer to the author/s of the texts attributed to Basil Valentine, regardless of whom he or they may have actually been.

Upon careful examination of *The XII Keys* and interpreting the substances and processes concealed and revealed therein, *The XII Keys* appears to be the work of a single alchemist who remained consistent throughout the treatise. The author of this specific Valentinian text however, oddly omits any therapeutic or medicinal use of the Philosophers' Stone anywhere in *The XII Keys*. The text appears to be purely procedural, including the final chapter on gold making. Other writings that support *The XII Keys* such as *An Elucidation of The XII Keys*, *The Manual Operations of Basil Valentine* and others included in *The Last Will and Testament of Basil Valentine* reveal processes that can be interpreted as pertaining to medicinal use of the Stone. However, this is valid as relating to *The XII Keys* only if Valentine's *powder of Gold of a purple colour* in the Phalaja recipes found in supporting texts indicates the Philosophers' Stone, rather than a literal interpretation that posits *powder of Gold of a purple colour* as the product of the *3rd Key*. This point is crucial because most works attributed to Basil Valentine are highly oriented towards medicinal or therapeutic applications. The notion that Valentine's *Phalaja Tincture* or *Stone* might perhaps be the missing *13th Key* to the therapeutic use of the Philosophers' Stone must also be considered.

The *Phalaja* conundrum presented in the *4th Key* is due to be expanded upon in an upcoming work entitled *Phalaja and the Fire Stone – Acetate Tincture Alchemy of Bacon and Valentine* by the same author.

INTRODUCTION

Keys to the Kingdom of Alchemy aims to decipher the chemistry encoded within each Key or chapter, and its corresponding imagery. I hope to demonstrate that whoever wrote *The XII Keys* was familiar with the archetypal recipe to confect the *Great Stone of the Ancient Philosophers* as interpreted here. Each Key guides the reader through a tutorial step for creating the Philosophers' Stone, as Valentine states the case to be:

> For a Co[n]clusion of these things, *I tell you for a truth, that one work proceedeth from another* …
>
> … hereunto minister a certain measure, and *mark my sixth Key, then proceed in the begun process, according to the order of the seventh, eighth, ninth, and tenth Key's* … go on with it to the appearance of the Kings honour and glory, to his highest purple garment …

The interpretation presented herein draws two conclusions that run contrary to current research: 1) the 2nd *Key* encrypts the identity of two *Hermes-Mercuries*, one apparent and the other hidden, and 2) that the *Salt of Ashes* that is the subject of the 4th *Key* is in accordance with archetypal Alexandrian alchemy. Indeed, something about the 4th *Key* has prevented many researchers from proceeding beyond that invisible barrier for centuries. I approached *The XII Keys* from the standpoint that the author was describing reproducible al/chemistry that should, if interpreted correctly, yield a product that qualifies as the Philosophers' Stone based on the archetypal Alexandrian model. This is a challenging notion to begin with because Valentine is most known for his emphasis on acetate oil and tincture alchemy and therapeutic applications. A careful reading of Valentine's works however, reveals that was a master antimonialist, expert at gold purification and particle size reduction, possessed at least three recipes for creating *butter of antimony*, could convert it to antimony oxychloride and oxysulfate and that he not only knew the basic mineral acids, but also employed various alchemical menstrua in unique ways. In short, he had more than the required knowledge and skill-set needed to create the archetypal Philosophers'

Stone. He apparently did not stop there: If the author of *The XII Keys* and supporting texts were indeed the same person/s, it suggests that Valentine advanced alchemical understanding by creating what could easily be described as new and improved versions of the Philosophers' Stone, including inventive methods for applying them medicinally. He created a contemporary Stone based on advanced work with antimony called the *Fire Stone*, yet this was not entirely original, being derived largely from Roger Bacon's antimonial Stone, *Lapidus Stibii*. Valentine's *Stone of Phalaja* (Pelaiah; פלאיה) is either pioneering new type of *Stone*, reflects Islamic salt, ash and distilled solvent methodology, or was a creative technique for converting the Philosophers' Stone to a more therapeutic or medicinal form.

Valentine was an innovative experimentalist and what truly sets him apart is that although he appears to value the archetypal alchemy of earlier traditions, he was not afraid to alter or improve upon known processes and thereby create new alchemical products. Valentine's work on *Phalaja* and the *Fire Stone* provided him with a wonderful template methodology for developing a variety of new product possibilities that can be considered authentic Philosophers' Stone variants by affecting a genuine union between gold and antimony in completely unique and novel ways. The goal of this study is to establish the fact that Valentine possessed a thorough grasp of archetypal Alexandrian alchemy and that this is what he was attempting to communicate to adepts and aspirants via *The XII Keys*. In typical alchemical fashion, the text was written for other adept alchemists who would have easily deciphered the substances and processes encoded in alchemical, religious, mythological and artistic symbolism.

How the text was written reveals much about the author/s. The author of *The XII Keys* fits the profile of a highly educated man, persistent to the point of obsession as regards product optimization (perhaps at the expense of process efficiency), well-versed in Judeo-Christian mysticism,

INTRODUCTION

Greco-Roman mythology, and seminal alchemical works of Morienus, Roger Bacon and Paracelsus. Apart from his native German, he likely spoke Latin and perhaps even Greek or Hebrew. The incredible result that illumines *The XII Keys* is that Valentine, whomever the author/s may have been, made masterful use of allegory and conceptual blending. Such an education, material costs, a functional laboratory and the leisure time necessary to pursue alchemical interests make sense if Valentine were indeed the Dominican friar as claimed, yet these would apply equally to a highly educated wealthy patron of alchemy such as Johann Thölde, or an eccentric German count or baron who valued anonymity.

I provide commentary, interpretation and suggest shortcuts for each Key's reproducibility that may bruise traditionalist sensitivities, but I do so only in the interest of research. For example, the result of the gold product at the end of the 3rd *Key* is described as a fine red powdered gold that becomes purple in color. The recipe to achieve this product is derived from chloroauric acid, the understanding of which provides earnest researchers with a few options such as purchasing chloroauric acid and evaporating it according to Valentine's instructions, thermally decomposing gold trichloride in a calcining dish or sourcing gold nanoparticles as a reagent and beginning at the 4th *Key*. Assuming that Valentine was skilled enough to achieve a relatively high level of purity as regard his reagents, a modern analog can be achieved by the use of commercial counterparts. The only difficulty with this approach is the product of the 4th *Key*, which requires glass of antimony – the crux of operative alchemical reproduction.

As is the case with any Philosophers' Stone recipe, artisanally preparing one's reagents is the most demanding aspect of the work. Reproducing the entire work in an artisanal manner strictly according to Valentine's text can be very rewarding. In the interest of modern research however, the experiment can be performed by bypassing redundant, outmoded or inefficient processes with good results.

BOOK 1

Preparing the Tria-Prima

*Soul, Spirit, and Body.
These are the three prime principles,
which in a coagulation come to a Mercury, Sulphur and Salt,
these three being in conjunction, according to the nature of the seed
produce a perfect body, be it in the Kingdom,
either of Minerals, Animals, or Vegetables.*

Basil Valentine

Key I - Grey Wolf and Lion King

... make a great Fire, into which cast the Wolf,
that he be quite burned,
then will the King be at liberty again ...
then hath the Lion overcome the Wolf ...

Basil Valentine

Basil Valentine's *1st Key* addresses gold refining methods in use at the time of his writing. These include gold parting using stibnite and/or mineral acids. The objective was to refine gold to the highest purity possible. Stibnite was known to post-medieval alchemists as the *Grey Wolf*, and gold as the *King*, *Lion* or the *Sun*:

> *Know, my Friend, that impure and defiled things are not fit for our work ... that which is good is hindered by that which is impure.*

> *... ought bodies to be purged and purified from all their impurities ... Our Masters require a pure and undefiled body [gold], which is not adulterated with any spot or strange mixture ...*

> 1. *... take the most ravenous grey Wolf... he is the Son of old Saturn ...;*
> 2. *cast unto him the King's body ... ;*
> 3. *make a great Fire, into which cast the Wolf, that he be quite burned, then will the King be at liberty again : When you have done this thrice, then hath the Lion overcome the Wolf ...*

> *... when the Lion is satisfied his spirit is made stronger than it was before, and his Eyes shine with great splendor like the Sun ...*

The instructions above clearly describe gold purification via stibnite followed by cupellation. Gold dissolves readily into molten stibnite.

BOOK 1 – PREPARING THE TRIA–PRIMA

Valentine apparently allows this alloy to cool, because the next step involves casting the *grey Wolf* into a great fire to vaporize the antimony.

The word *burned* used by Valentine can be better understood as *vaporized* via cupellation at a temperature lower than the melt temperature of gold but high enough to remove lead oxides (pictured in the Prima Clavis woodblock print) or to vaporize quicksilver (mentioned in supporting documents). The process is repeated twice more before the *Lion* (gold) shines like the *Sun* (is purified) and is fit for further alchemical treatments.

Supporting Documents

In *The Manual Operations of Basil Valentine*, the anonymous author presents the same process in detail with the introduction of quicksilver, revealing that the final product should be a fine gold powder:

1. Take of the very best Gold you can have one part,
2. of good Hungarian Antimony six parts,
3. melt this together upon a fire, and pour it out into such a pot as the Goldsmiths use; when you have poured it out it becometh a Regulus.
4. This same Regulus must be melted again, that the Antimony may be separated from it.
5. This being done, add to it [philosophers'] Mercury, and melt it again, and cleanse it again.
6. Repeat this the third time; and the Gold is purged and purified enough for the beginning of the Work.
7. Then beat the Gold very thin, as Goldsmiths do, when they gild [gold leaf], and make an Amalgama with common Quick-silver, which must be squeezed through a Leather;
8. let the Quick-silver, fume away by little and little upon a gentle fire, that nothing of it may remain with the Gold, and stir it about continually with a small Iron, and the Gold is become subtile, so that its water may the better work upon it and open it.

Irrespective of the addition of quicksilver, or whether the two texts were written by the same author, the objective was a purified fine gold powder achieved by the two-step process of antimony purification and either lead or quicksilver cupellation.

Valentine addresses gold parting using mineral-acids later in the chapter as a chemical alternative to the metallurgical, highlighting the importance of removing any remaining traces of acid from the gold.

> *If also by a medium a corrosive [acid] should be joyned, by which our body [...of the King; gold] might be dissolved, see that all the corrosive be washed away ...*

Valentine likens quartation to grafting good fruit trees to poor ones in order to produce a better tree with more pleasing fruit. In such circumstances, he would alloy 3 parts or more of silver with 1 part gold and then remove the silver with nitric acid. He then explains in figurative language that the King (gold) will feature in the six planetary stages of the chemical reaction and finish in the seventh. This is another way of stating that he begins with gold and the resulting Philosophers' Stone is gold in its highest exaltation, figuratively as a palace (the reaction vessel) being *adorned with golden tapestry* (the Stone). During antiquity, the word *Ios* (tincture) meant purple, suggestive of the royal purple robe initially worn by Persian kings.

> *The King walketh through six places in the Celestial Firmament, but in the seventh he keep[s] his seat; for the King's Palace is adorned with golden Tapestry: If now you understand what I say, then have you opened the first Lock with this Key ...*

The imagery depicted in the *Prima Clavis* is rather straightforward, and each detail tells a specific story when understood from the perspective of operative alchemy. He titled the picture the *Prime Key* because it represents the two prime materials – gold and antimony, as King and

BOOK 1 – PREPARING THE TRIA–PRIMA

Queen – the Red Man (Apollo; gold) and his White Wife (Diana; antimony) and the King's Palace necessary to create the Philosophers' Stone.

Artisanal gold refining as depicted in Valentine's *Prima Clavis* can be a fun experiment but is relatively unnecessary in light of the fact that high-purity gold is readily available today in the form of 99.99% pure edible gold leaf and powders. Once gold is fully purified and converted to a powder or calx, it must then be dissolved by special solvents, two of which feature in Basil Valentine's *2nd Key*.

KEY I – GREY WOLF AND LION KING

Symbolism of Prima Clavis

- **King's Palace in left background** = the Grand Edifice of Alchemy by which the King's Power is brought forth
- **King taking a step** = gold purification as the first step (according to Valentine)
- **Queen with flowers** = calcined antimony as flowers of antimony
- **Queens peacock feather whisk** = various colors antimony can assume
- **Grey wolf leaping a crucible** = gold parting via stibnite (old Saturn's son)
- **Old Saturn** = cupellation to achieve highly purified gold powder (killed gold)

Key II – Two Mercuries

And this is the Philosophers Mercury, or Mercurius Duplicatus, and are two spirits, or a spirit and water of the Salt of Metals.

Basil Valentine

B asil Valentine's *2nd Key* encrypts the recipe for two very important alchemical solvents – one common strong acid and another secret mineral solvent crucial to all forms of alchemy based on Alexandrian archetypal methods for creating the Philosophers' Stone. The first sentence of the chapter encrypts the message that he is describing more than one kind of *drink*, each with its own potential.

> In the Court of great Potentates *various kinds of drink are found*, yet scarce any of them alike in smell, color, and tast[e], for *their preparation is different*; yet all they all drink, because they are all made and *necessary for their particular uses* in the family. ...

1st Mercury – The King's Bath

He explains that when the *Sun* (gold), shines (is purified), that it attracts water (acid) causing it to rain (precipitate gold). His first acid is aqua regia created from *Niter* and *Sol Harmoniack*. Niter was also known as Saltpeter or Sal Petra to European alchemists, the chemical identity of which is potassium nitrate (KNO_3). *Sol Harmoniack,* or by its more common spelling *sal ammoniac*, is a natural salt derived primarily from urine deposits during antiquity, its refined chemical identity being ammonium chloride (NH_4Cl). A passage from a supporting text, *An Elucidation of the XII Keys*, clearly states that this is the case:

> ... my second Key informeth thee of, namely *what matter you ought to take to the* Kings Balneum *[or King's Bath; aqua-regia], whereby the*

BOOK 1 – PREPARING THE TRIA-PRIMA

> King *is destroyed, and its external form broken*, *and its undefiled* Soul *may come forth, to this purpose will serve the* Dragon *and the* Eagle, *which is* Niter *and* Sol Harmoniack, *both which after their union are made into a* Aquafort *[strong acid] ... into which* Balneum *the King is thrown ...*

Valentine's equation for the obvious or apparent *Mercury I* in the language of alchemy and chemistry is as follows:

Dragon-Serpent + Eagle-Bird of Hermes = Hermes / Mercury I
Niter + Sol Harmoniack = Hermes / Mercury I
$NH_4Cl + KNO_3 = NH_4NO_3 + KCl$

Supporting Documents

The precise recipe for creating the *Aquafort* occurs in a supporting text, *The Manual of Operations of Basil Valentine*:

1. Take *one part of Salt-peter* well purified, and grind with it *the like quantity of Sal-armoniac, and* half as much Pebbles *[either boiling stones or green vitriol] very well cleansed and washed.*
2. Mingle all these ingredients together, *and* put them into an Earthen Retort, *that the Spirits may not come through [leak; escape], put the same into a Distilling Furnace: The Retort must have a Pipe behind and* put as large a Receiver as you can get to the Retort.
3. The Receiver must lye in a vessel full of cold water, *and a wet Linnin-cloth must be put round about it, which you must* wet continually *with another wet Cloth;* then put again so much Matter into the Retort till all is gone into it, and then your Water is prepared.

Lunar Principle (serpent) = Niter
Solar Principle (eagle) = Sol Harmoniack

Here Valentine is creating an ancient form of aqua regia attributed to Islamic alchemy. He then gives instructions for gold's dissolution in the same lecture. A key note in what follows is that either Valentine uses a *Calx of Gold*, suggesting that when he purified his gold according to the

1st Key, he employed lead or volatized quicksilver in a cupel without bringing gold to a molten state, thus the gold product remained in powdered form known to alchemists as *Calx of Gold*:

1. *Take then of the prepared Calx of Gold one part*, put it into a Glass body,
2. and *pour three parts of the above made Water upon it*, and place it in warm Ashes, and the Gold will dissolve in it;
3. but if it should not altogether be dissolved pour more fresh water upon it, and it will dissolve all.

The process of dissolving the *Calx of Gold* thus finished, he then uses a very complex method of distilling to remove the liquid from the solids leaving *Fixed Powder of Gold* remaining, the chemical identity of which is likely tetraaminegold nitrate ($[Au(NH_3)_4](NO_3)_3$). This is of a similar species to gold fulminate (explosive gold), known today as gold hydrazide. Valentine's version however, was more stable and heat resistant than other types of fulminating gold. The hypothesized chemistry for Basil Valentine's 1st Mercury, the *King's Bath* is as follows:

$$32\ Au + 38\ NH_4Cl + 27\ KNO_3$$
$$= 11\ HCl + 8\ [Au(NH_3)]_4(NO_3)_3 + 27\ KCl + 9\ NH_4OH$$

A shortcut is to add ammonium chloride to nitric acid to dissolve gold:

$$32\ Au + 38\ NH_4Cl + 27\ HNO_3$$
$$= 38\ HCl + 8\ [Au(NH_3)]_4(NO_3)_3 + 9\ NH_4OH$$

2nd *Mercury – The King's Palace*

He then segues into a hint regarding reagents required to create his 2nd Mercury.

For the building of a Princely Palace, various and divers[e] Workmen and Mechanicks must be set on work, before it be called a beautiful & perfect

> *Palace. Where* stones [mineral-based reagents] are required, *wood [plant-based reagents] must not be used. ...*
>
> *When thus* the Kings Palace is prepared and adorned by several workmen, and the glassy Sea is finished, *and the Palace furnished with goods, then may the King safely enter and keep there his Residence.*

According to Valentine, the *King* (gold) must first be purified, and then he must *bathe* (in aqua regia), and the *glassy sea must be finished* (gold-antimony glass) before he can enter his *Palace*. *Mercury II* was known in Alexandrian alchemy by the cover-names *Divine Water, Hermes, Pharmakon* and *Eudica* among others and it is indeed a *water* created from a mineral-based reagent – powdered *flowers* or *regulus of antimony* distilled with sal ammoniac.

One of the ingredients is already cleverly encrypted by Basil Valentine by the cover-term *Sol Harmoniack*. Valentine used the word *Sol* to indicate a salt associated with the sun. Sal ammoniac is symbolized by an asterisk, but has traditionally been considered a solar salt as well as a stellar one. The mis/spelling of *Harmoniack* was intended to suggest attraction, opening and establishing harmony, the salt being crucial to synthesizing each of the two *Mercuries* thus it finds harmony with each and is the uniting principle resulting in a double salt. Its production is likened to a marriage between Apollo, god of light and the sun, and Diana, goddess of the moon and birth. These reagents must be *naked* (purified) prior to the reaction:

> *... when the Spouse is to take a carnal cognizance of her Husband,* all these Garments are laid aside, neither doth she keep any thing on her, *but what the Creator granted her at the beginning.*
>
> *Even so our Bridegroom Apollo, with his Bride Diana, is to be married ... But my friend know, that* the naked Bridegroom must be espoused to his naked Bride *... that* they may lye down as naked as they were born, that their seed be not destroyed *by any strange Mixture.*

Valentine's equation for the hidden *Mercury II* in the language of alchemy and chemistry is as follows:

Dragon-Serpent + Eagle-Bird of Hermes = Hermes / Mercury II
Diana + Apollo = Hermes / Mercury II
$Sb_2O_3 + NH_4Cl = SbCl_3$

> *For a conclusion of this discourse, I tell you truly, that the most precious water, wherewith the Bridegrooms Bable [tower; palace] must be made, must be wisely and with great care prepared of two Fencers (understand of the two contrary matters [solar and lunar]) that one adversary may drive out the other ...*

Lunar Principle (Diana) = Antimony Flowers-Glass
Solar Principle (Apollo) = Sol Harmoniack

The summary makes it perfectly clear that Valentine is indeed discussing a chemical reaction in terms that apply to both *Mercuries*:

> *... the Prize must be won: For what advantage is it for the Eagle, to build her Nest in the Rocks, where her Chickens will dye on the tops of the Mountains, by reason of the coldness of the Snow?*

> *But if you adde to the Eagle the old Dragon [antimony], which hath a long time had his habitation among Stones [vitriols], and creepeth out of the Caves [mines], and put them both in the Infernal Pit [distillation vessel], the[n] will Pluto breath[e] upon them [heat], and will enforce a fiery volatile spirit [distillate] out of the cold Dragon, which by its great heat burneth the Eagles feathers, and maketh a sweating [corrosive] Bath, that the Snow [Niter and/or Flowers of Antimony] on the highest Mountains melteth, and turneth into water [twin Mercuries; solvent/s].*

> *Whereby the Mineral Bath is well prepared, which bringeth riches and health to the King.*

When distilling *butter of antimony*, the product can crystalize in the neck of the condenser (snow on the highest Mountains), which must be melted (melteth) allowing the liquid distillate (turneth into water) to descend into the receiver. Whether the above is read from the perspective of operations to distil *aqua regia*, or *butter of antimony*, the passage applies to both. This is precisely why Basil Valentine's *2nd Key* is so crucial to confecting the Philosophers' Stone. It cleverly addresses both *Mercuries*, one common and apparent, while the other remains hidden or occult. *Mercury* is the Latin name for the Greek god *Hermes* – in great antiquity the secret solvent, or *Philosophers' Mercury*, was originally called *Hermes*. An interpretive look at the woodblock image of Basil Valentine's *2nd Key* (*Twin* or *Double Key*) can unlock the secret of the twin Mercuries.

Basil Valentine's *2nd Key* is a masterpiece of encryption. His *1st Mercury* allowed him to dissolve gold to a *Fixed Powder of Gold*. His *2nd Mercury* was such a powerful secret that its identity was carefully guarded by every adept-alchemist in possession of it, thus it remained the *hidden mercury*. His *2nd Mercury*, also known to alchemists as *Philosophers' Mercury* or *Sophic Mercury,* is a nonaqueous liquid antimonial solvent that can dissolve gold without violence and then coagulate it to a salt or glass.

Since the earliest Alexandrian alchemists began using it, *butter of antimony* has remained central to recognizable alchemy throughout Alexandrian, Islamic and European alchemical traditions. This solvent is the subject of the *5th Key*, achieved by at least three different recipes that occur in various works attributed to Valentine. Knowledge of antimony oxysulfate, trichloride and oxychloride suggests that Basil Valentine was fully familiar with the archetypal method for creating the Philosophers' Stone, which is explained in detail in *Cracking the Philosophers' Stone* by the same author. As we will discover, he actually improved the archetypal process by reducing gold to a finer state than had ever been achieved before. The secret to his gold crystallization and particle-size reduction however, is encrypted in his *3rd Key*.

KEY II – TWO MERCURIES

Symbolism of II Clavis

- **Mercurius Duplicatus** = Aqua Regia and / or Butter of Antimony holding two caducei
- **Exposed swordsman** = Islamic Aqua Regia
- **Hidden swordsman** = Butter of Antimony (the hidden Key and great secret of alchemy)
- **Serpent-Dragon** = Niter and / or Antimony (antimony is portrayed as royalty; one of tria-prima)
- **Eagle-Bird of Hermes** = Sal ammoniac salts of Augila (Eagle), 2nd stop on the trans-Saharan caravan route according to Herodotus
- **Solar principle** = right swordsman looking at the sun / masculine fleur-de-lis on right caduceus
- **Lunar principle** = left swordsman gazing at moon / feminine fleur-de-lis on left caduceus
- **Sword/s** = chemical reagent/s
- **Hidden tower in background** = tower of the temple of Amun in Libya (common in alchemical imagery)
- **II Clavis** = the double key hiding twin recipes for double Mercuries (Mercurius Duplicatus)

Key III – Cœlestial Sulphur

... thus is the Philosophers Sulphur well prepared for your work, and this is the Purple Mantle, or Philosophick Gold ...

Basil Valentine

Basil Valentine's *3rd Key* encrypts the recipe for two very important alchemical forms of gold – gold trichloride salt crystals and gold nanoparticle powder. The *fiery Sulphur* below, so named likely because it was a form of explosive gold, is the *Fixed Powder of Gold* from the *2nd Key*. Fiery Sulphur must be dissolved in another type of *water*, or *solvent* and converted to *Cœlestial Sulphur* as alluded to by the following:

> By water fire may be wholly extinguished, if much water be cast into little fire, then the fire gives way to the water, and yeildeth up the victory unto it: So must *our fiery Sulphur be conquered, and overcome by water prepared according to art*.

Supporting Documents

The exact recipe for this third solvent, or *water prepared according to art*, is found in a supporting text, *An Elucidation of the XII Keys*:

1. Then take of good *sprit of Salt-niter [nitric acid] one part*, and of dephlegmed *spirit of ordinary Salt [hydrochloric acid], three parts*,
2. *pour these spirits* together warm'd a little, into a body *on the fore-written Gold powder*, lute a Helmet and Receiver to it,
3. *drive the gold over* as formerly in sand several times *with an iterated distillation*, the oftner the better, let the Gold come to be volatile more and more and at last let it all come over.

Valentine explains that the purpose is to divide and open the gold, which can be understood today as an attempt to reduce the particle size of the gold clusters to smaller clusters. He then explains that after fully volatizing the gold, the liquid must be removed via gentle evaporation (separation of the water) – a recrystallizaition process to remove excess hydrogen chloride and to achieve relatively pure gold trichloride salt crystals. This is described in the passage from *The XII Keys* below:

> *If after the separation of the water, the fiery life of our Sulphureous Vapor can but again triumph and obtain the victory; but no conquest can be herein obtained, unless* the King adde force and power to his Water, *and hath* given it the Key of his own proper colour, *that* he may be thereby destroyed and made invisible; *yet at this time* his visible form ought to return, *yet with a diminution of his simple Essence, and melioration of his Condition.*

The process is stated more clearly in a parallel passage from *An Elucidation of the XII Keys*:

> *Note further, that after this work* those salt spirits [the acid] must be abstracted from the Gold, *which was driven over, very gently in Balneo Marie,* let nothing of the tincture of the Gold come over ...

At this stage of the process presented in the *3rd Key*, Valentine has achieved gold trichloride crystals ($AuCl_3$), yet they are still corrosive and need further refinement. Valentine is trying to achieve an incombustible form of high-purity finely divided gold powder. It appears as though Valentine is attempting to improve upon the basic types of *Gold Calx* common to alchemy. Common *Gold Calx* was gold divided by repeated mercury distillations, yet this process will divide gold only just so far, resulting in a product known synonymously as *Gold Calx*, *Gold Lime* or *Gold Sponge*. The other common *Gold Calx* was explosive *Gold Fulminate* powder briefly mentioned in the previous chapter. Valentine appears to have been dissatisfied with either standard version of *Gold Calx* and innovatively dissolves his *Fixed Powder of Gold* in the type of aqua regia still used in modern chemistry labs today. The problem as perceived by

Valentine was that his process yielded gold-salt crystals rather than a fine gold powder.

> *So he that would prepare incombustible [Cœlestial] Sulphur of the Philosophers, let him first consider with himself, that he seek our Sulphur in that wherein it is incombustible, which cannot be, unless the Salt Sea have swallowed up the body, and cast it up again : Then exalt it in its degree, that it far exceed in brightness all the other Stars in the Heaven.*

According to supporting documents, he solved this problem by a simple technique he called *exaltation*, which apparently meant simply warming the *Crystals of Gold* with gentle fire to thermally decompose the gold salts. The process is explained in detail in *An Elucidation of the XII Keys*:

1. ... take that Gold, or rather these *Crystals of Gold, from which you have separated the water*, put it in a Reverberating pan, set it under a muffle [furnace],
2. let its first fire be gentle for an hour, *let all its corrosiveness be taken away*, then *your powder will be of a fair scarlet color, as subtle as ever was seen* ...

Chloroauric acid is golden yellow initially, but changes to red and finally purple when heated and moisture is removed. This does not appear to be what Valentine was referring to when he instructs the operator to *let all its corrosiveness be taken away* because red gold trichloride is still very corrosive, sometimes called *acid gold trichloride*. When *gold trichloride* is heated to around 160 °C, it decomposes to *gold chloride*.

$$AuCl_3 = AuCl + Cl_2$$

When *gold chloride* is heated to 298 °C, it decomposes it to a very pure *gold-calx*. Unfortunately, the answer to Valentine's scarlet colored calx is not quite as simple as simple thermal decomposition.

$$AuCl + \Delta = Au + HCl$$

BOOK 1 – PREPARING THE TRIA-PRIMA

His color indicators may initially suggest red gold trichloride crystals dissolved and purple gold hydroxide precipitated therefrom. If Valentine's description of a subtle scarlet non-corrosive gold powder is accurate however, then his is the first clear indication of achieving dry gold nanoparticles in the history of European alchemy, which reveals tremendous effort exerted towards dividing gold's particle size. Scarlet red coloration suggests a particle size somewhere between 5–24 nanometers, yet later he clearly describes his powder as being purple, which suggests aggregation into the 24-90 nanometer size range. In either case, pending reproducibility, Valentine deserves credit as the first European to achieve dry gold nanoparticles and colloidal gold in solution.

While gold in this form would be more than adequate for most alchemists, Valentine took his product to yet a further stage of *exaltation* according to supporting texts. The process is detailed in *An Elucidation of the XII Keys*:

> ... put it [the concentrated scarlet gold-salt solution] in a clean viol, pour on it fresh spirit of ordinary Salt [hydrochloric acid], first brought to a sweetness [with ethanol], let it stand in gentle digestion, let that spirt be deeply ting'd and transparent, red like a ruby [colloidal gold solution], [de]cant it off, pour on fresh, extract again, iterate the work of canting and pouring on, till no more tincture of it appeareth, put all these extractions together, separate them in Balneo [water bath] gently from the Sulphur of Sol, then that powder is subtle and tender, of great worth ...

... and again in *A Brief Appendix And plain Repetition or Reiteration of Basil Valentine*, yet without mention of the separation step:

> ... if the spirit of common salt be united with spirit of Wine, and both be three times distilled over together, then it waxeth sweet, and looseth its acrimony: This prepared spirit doth not corporally dissolve Gold, if it be poured on a prepared Calx of Gold, it extracteth its highest tincture and redness ...

KEY III – CŒLESTIAL SULPHUR

Valentine's technique of employing a combination of hydrochloric acid + ethanol as a means of volatizing his gold was mirrored during the 17th century by none other than the Honorable Robert Boyle, who employed a virtually identical process to create his version of *Aurum Potabile*, preserved in *The Philosophical Works of Robert Boyle, volume 1, page 63*:

> And Farther, to make *a preparation wherein gold is reduc'd to very minute parts*, without the help of mercury, or of any precipitation by means of sharp salts,
>
> 1. we took *refin'd gold, and dissolv'd it in clean Aqua regia**;
> 2. we first, *with a very moderate heat, drew off the superfluous liquor, whereby the gold, with the remaining part of the menstruum*, was left in the appearance of *a thick oily liquor* [Valentine's Scarlet Sulphur of Sol].
> 3. This done, *we pour'd upon it a treble weight of vinous spirit* [ethanol], *totally inflammable; and in a short time we had a very subtile powder, or high-colour'd calx of gold*, that subsided at the bottom; *the menstruum being strangely dulcify'd* [sweetened] *as to tast[e], and become fragrant in point of smell.*
> 4. *A very few days after, we decanted the liquor, and put on it fresh ardent spirit;* when leaving them a while together, *there subsided the like well-colour'd calx*, more plentifully than before; whilst *the menstruum acquir'd such qualities as made it seem likely to prove an useful medicine.*
> 5. *This powder of gold*, tho' it seem'd not to require it, *we further purified for internal use, by burning a totally ardent vinous spirit, twice or thrice thereon, to carry off with it any little corrosive* or saline particles, that might have still adher'd to the metalline ones.
>
> * *The spirituous Aqua regia*, here mention'd, *which may probably be a more innocent menstruum in preparing gold for medicinal uses*, I very easily make, by mixing *one part of good spirit of salt with two of strong spirit of nitre.*

The primary difference between Valentine's process and Boyle's is that Valentine extracted all of the gold into solution, and then separated the liquid from the solids in a water-bath, whereas Boyle decanted the liquid, retaining it as a potential medicine. In both cases, the result is an extremely fine *Calx of Gold* derived from concentrated gold trichloride solution.

According to the supporting materials, Valentine created an ether-like substance by distilling hydrochloric acid and ethanol – the chemical identity of which is crude ethyl chloride, a.k.a. chloroethane but known originally as hydrochloric ether. Paracelsus was also familiar with ether, being one of the earliest Europeans to work with it. In supporting texts attributed to Valentine in the two passages above, the procedure involves either dissolving gold trichloride crystals or suspending gold nanoparticles in crude chloroethane solution, then separating the liquid from the finely divided gold by gentle evaporation. Valentine and Boyle's processes were improved upon during the 18th century, serving as precursor to the use of diethyl ether to create *Aurum Potabile*. Anecdotal and circumstantial evidence suggests that a process such as this was a method later employed by the Comte de St. Germain.

Two centuries after Valentine, the process still fascinated chemists and physicians such as Dr. Frobenius, F. R. S. of London, who wrote of it in *An Account of a Spiritus Vini Æthereus*, with experiments performed for Robert Boyle in Boyle's laboratory and compared to similar experiments conducted by Sir Isaac Newton (published after Boyle's death 1st of January, 1729):

> And indeed *a wonderful Harmony is observable betwixt Gold and this Æther, even greater than between Gold and ℞Regia; insomuch as from hence Gold appears to approach nearer to the Nature of Oils than of Earths, as shall be proved when we treat in their proper Place of the three harmonious Menstrua which we have discovered … If a Piece of Gold be dissolved in the best Aq. Regia, and upon the Solution Cold, be poured half an Ounce, or what Quantity you please of the Æthereal Liquor, shake the Glass carefully, and all the Gold will pass into the Æthereal Liquor … The Æther will swim like Oil on the surface of the*

KEY III – CŒLESTIAL SULPHUR

> *corrosive Waters. The Experiment deserves the utmost Attention; for here the heaviest of all Bodies, Gold, is attracted by this very light Æther …*

The process described above is the volatization of gold that fascinated European alchemists, which is to say the lightest liquid attracts and suspends the heaviest solid (as it was understood by alchemists and early-modern chemists). Regardless whether Valentine reduced gold trichloride via thermal decomposition or via tincture and evaporation, it is safe to say that the desired product of the *3rd Key* is gold powder at the finest state of division possible according to gold technology of the period. If the process described by Valentine was indeed the chloroethane tincture followed by evaporation, it suggests Valentine may have been the first European to achieve stabilized dry gold nanoparticles. The Philosophers' Stone is composed of three primary ingredients – *Astral Salt, Sulphurous Soul* and *Mercurial Spirit*. The *Sulphurous Soul* principle in Valentine's work was precisely this *Essence of Blood* as red-purple gold.

> *And in its own Essence is so full of blood … This is the Rose of our Masters, of a purple color, and the Red Blood of the Dragon, whereof so many have written ; it is that purple Mantle … Keep safely this honourable Mantle, together with the Astral Salt [of the IV Key], which followeth this Cœlestial Sulphur [of the III Key] … and give unto it of the volatility of the Bird [of the V Key], as much as will suffice …*

Valentine refers to his highly refined gold by the following cover-terms:

- Honorable Mantle
- Rose of the Masters
- Red Blood of the Dragon
- Purple Mantle
- Cœlestial Sulphur

Basil Valentine's *3rd Key* cleverly describes the process of dissolving *Fixed Powder of Gold* (a form of typical gold calx) in the *Salt Sea* (aqua regia), by which he achieves *Crystals of Gold, fiery Sulphur* (pictured left in aqueous state). This product is then either thermally decomposed to *Cœlestial Sulphur* (pictured right) or converted to an extremely fine powder via tincture and evaporation. He leaves absolutely no doubt up to this point that he is describing a precise pedagogical method for preparing his reagents to a very high standard of purification.

Today researchers have several options for reproducing the product of *The XII Keys*, while bypassing some of the more dangerous or lengthy processes. Shortcuts for the *3rd Key* might include:

1. Thermally decomposing chloroauric acid (pictured);
2. Create Valentine's *Sweet Spirit of Salt* (liquid chloroethane; ethyl chloride) and apply it to commercial gold trichloride or fine gold calx and separate via gentle evaporation;
3. Reproduce Boyle's powdered *Aurum Potabile* via his recipe
4. Apply the diethyl-ether process to chloroauric acid or gold trichloride crystals as described by Dr. Frobenius
5. Purchase gold nanoparticles, or medicinal Indian *Swarna Bhasma* as a ready-prepared gold ingredient

In any case, finely divided and highly pure gold is a necessary ingredient in the product of the *4th Key*.

The dragon featured in the foreground of the wood-block print is actually the product of the upcoming Key. For this *3rd Key* and the upcoming *4th* and *5th Keys*, Valentine includes artistic symbolism that links the current image to the one that follows. This is meant to highlight the importance of the *3rd*, *4th* and *5th Keys* as representative of the preparatory work –

creating *Sulfur*, *Salt* and *Mercury* principles – the three prime ingredients for confecting the Philosophers' Stone.

As elucidated in *Cracking the Philosophers' Stone*, during the industrial period of Alexandrian alchemy (prior to the 4th century), technology for purifying and particle-size reducing gold was rarely if ever an issue. This is because fine gold powders and dust at relatively high purity were readily available to Alexandrian artisans.

Notice the emphasis Valentine places on gold technology revealed in his first three Keys. He devotes 25% of his book to the processes of gold purification, chemically converting it to gold calx, dissolving and crystalizing it to gold salt followed by the salts' conversion to a nanoparticulate, or at the very least an extremely fine powdered form.

The use of chloroethane (or hydrochloric ether, a.k.a. ethyl chloride) and later diethyl ether and acetone, was to serve as a carrier of of gold salts. Gold in this form was typically described by alchemists as the *Sulphur of Gold*. The carrier of this "sulphur" in other alchemical traditions was one of the solvents mentioned above, all synonymously known as *Philosophical Mercury* in alchemical traditions that relied on salts, ashes, distillation and the use of mineral acids that typify late-Islamic and European alchemical traditions. This particular evolution of alchemical technique reflects later expressions of alchemy however, and is somewhat removed from the archetypal original due to the simple fact that mineral acids (apart from antimony trichloride) had not yet been discovered during the technological period of Alexandrian alchemical expression.

Both Basil Valentine and Paracelsus demonstrate in their work that they were familiar with both antimony-based and the evolved solvent-based styles of alchemy. We can observe a turning point and perhaps bifurcation of these two styles of alchemy in European processes for confecting the Philosophers' Stone where mineral acids, ether or acetone and organic salts were employed to confect an evolved form of the Philosophers' Stone. Valentine's work in general, and *The XII Keys* in particular has often been interpreted as reflective of this evolved expression, and overwhelming evidence indicates that he had mastered these types of alchemy as well, yet in his own words, he is adamant that the *Great Stone of Ancient Philosophers* (the archetypal antimony-based variety) is the topic being presented in *The XII Keys*.

Gold technology was a primary aspect of European alchemical traditions as evidenced by a shift in emphasis demonstrated by Valentine and others. Where Alexandrian alchemists began with stibnite calcination, European alchemists routinely began with gold purification and reduction. Basil Valentine understood his *Cœlestial Sulphur* to be one of the three primary ingredients (the *Tria Prima*), and a major component of the product described next in Valentine's 4th *Key*.

KEY III – CŒLESTIAL SULPHUR

Symbolism of III Clavis

- **Red rooster attacking** = European aqua regia attacking Fixed Powder of Gold
- **Red rooster being devoured** = European aqua regia being overcome via final evaporation and fixation
- **Red fox** = Fixed Powder of Gold attacked, fiery Sulfur, aka Crystals of Gold created and converted to Cœlestial Sulphur
- **Beast-Griffin** in foreground = Intended dry application of gold nanoparticles in the 4th Key
- **King's palace in center background** = the King's home in a different, nearer land than in the 1st Key
- **Rocky earthy terrain** = suggests dry earthy product of the 4th Key
- **Icy mountainous terrain** = in background suggests glacial oil of antimony product of the 5th Key

Key IV – Salt of Ashes

If the Artist want Ashes, he cannot make Salt for our Art,
for without Salt our work cannot be made into a body,
for Salt only coagulateth all things.

Basil Valentine

Basil Valentine's *4th Key* encrypts the recipe for a gold-antimony glass compound that he referred to as *Salt of Ashes* or *Salt of the Philosophers*, traditionally depicted as a monstrous griffin or type of dragon as was done in the woodblock image associated with the *3rd Key*. At this stage in the recipe for confecting the Philosophers' Stone, the process of creating two chemical compounds begins – the first being the *Salt of the Philosophers* addressed in this Key. In the opening passage, Valentine refers to ground antimony glass by the cover-name *Earth*, explaining that it must be "corrupted" with something that adds new life to it. The matter must then be turned to *Earth* again, in other words *ashes-to-ashes* yet infused with new life in its final form. The resulting revivified product is Valentine's *Salt* principle, referred to as *Salt of Ashes* or *Salt of Philosophers*.

> All flesh that came from the Earth, must be corrupted and return to Earth again, *as it was Earth at the first, then that Earthly Salt begetteth a new generation, by a Cœlestial revivification,* for if it were not first Earth [Prime Material; Adam], there could be no revivification in our work; for in the Earth is the Balsom of Nature, and is their Salt …

Cœlestial revivification suggests a type of enlivening using *Cœlestial Sulphur* as an ingredient, thus the product of the *3rd Key* finds application in the *4th* as a type of *seed* or *ferment*. *Cœlestial Sulphur* was considered the *Soul* of gold, and when embedded in antimony glass was thought to

BOOK 1 – PREPARING THE TRIA-PRIMA

add life to, or revivify, the new product referred to as *Salt*, *Glass* or *Lion's Blood* as explained in *An Elucidation of the XII Keys*:

> ... look upon the body of Gold, not as if no other benefit could be reaped of it, but only his Soul; not so: impute no such weakness into that body, but after you have drawn forth its [Cœlestial] Sulphur, there is yet in it [in potentio] the Salt of glory, and the triumphant victrix ... Therefore take your Solar earth, out of which you drew your seeds, or the true Lions bloud, and reduce it by reverberating to a fix'd powder, and subtile impalpable ashes, extract from thence a very subtile Salt as bright as Ivory ... the body of Sol is anatomized [dissected] by the particulars ...

The chemical identity for the *particulars* mentioned above, that aid in the further dissection of gold is a matrix antimony trioxide and trisulfide, also referred to by the cover-terms *Ashes* and *true and genuine Tartar*.

> ... that which was made by the Creator of nothing, must by Fire be burnt to Ashes ... For in those Ashes lye the true and genuine Tartar[us] which must be dissolved, and when that is dissolved, the strongest Lock of the King's Palace may be opened.

The substance hidden in the 4th *Key* is so well encrypted that very few have ever unlocked its secret, which thwarted understanding of subsequent Keys. Practical reasons exist for these difficulties, the primary being a misinterpretation of *Salt of Ashes*. Common salt of ashes, aka *Philosophers' Tartar*, is indeed a salt extracted and crystalized from burnt ashes, its chemical identity being a potassium salt. *Philosophers' Tartar* is a very common substance used for a number of processes in plant alchemy, but has little to do with archetypal alchemy.

It would be premature to conclude that Valentine was referring to common *tartar* without bearing in mind his fluency with mythological representation. According to Hesiod, Tartarus was one of the primordial deities following Chaos and Gaia, yet preceding Eros. In Orphic sources and in the Mystery schools, Tartarus was the unbounded *first existent*

KEY IV – SALT OF ASHES

(Primum-ens) from which light and the cosmos originated. In Greek mythology, Tartarus was associated with both deity and the underworld. Valentine would have been aware of these associations when he employed the cover-name *Tartarus*.

Supporting Documents

In a commentary found in the Valentine writings, the recipe for the *Phalaja* products appears identical to *The XII Keys*, this notion resting entirely on the chemical identity of *Salt of Ashes*, described in the *Phalaja* commentary as a potassium salt. This creates an alchemical conundrum because if the interpretation herein is correct, then the question as to whether the author of *Phalaja Tincture / Stone* also authored *The XII Keys* must be addressed. At least three scenarios result from this conundrum:

1. **The author of *The XII Keys* and Phalaja recipes is the same author** – in this scenario, *Salt of Ashes* is in reference to antimony ashes, antimony trioxide, fused with gold according to the archetypal recipe. By this interpretation, *The XII Keys* is in accordance with archetypal Alexandrian alchemy. *Phalaja* is a unique separate substance based on the gold technology of Keys 1-3, *salt of ashes* in the *Phalaja* application being a potassium salt and is completely unrelated to the product of the 4th Key.

   An alternate reading is that the *Phalaja* work is based on the Philosophers' Stone as the identity of *powdered Gold of a purple colour* as the main reagent in a process that makes use of potassium salt where again *salt of ashes* in the *Phalaja* recipes remain unrelated to the 4th Key. In either of these scenarios, *The XII Keys* and the supporting texts were written by the same highly innovative and experimental alchemist/s, yet the *Salt of Ashes* in the 4th Key and potassium salt in the *Phalaja* recipe remain unrelated. The recipe in *The XII Keys* and the *Phalaja* recipes are similar in procedure and terminology, yet different as regard substances.

2. **The author of *The XII Keys* and the author of the *Phalaja* work were different writers** – in which case an alchemist-commentator attempted reproducibility of *The XII Keys* and erroneously deciphered *Salt of Ashes* in the 4th *Key* as potassium salt typical of plant-alchemy, a misinterpretation thus leading to the new and truly original *Phalaja* products. The misdirected author realized that he failed to reproduce the authentic *Great Stone of the Ancient Philosophers* and thus gave his product a new name. In this scenario, the interpretation herein remains valid and it would not be the last time that the 4th *Key* misdirected or baffled alchemical researchers.

3. ***Salt of Ashes* in *The XII Keys* refers to potassium salt** – in which case the author of *The XII Keys* and the *Phalaja* products is the same person and *The XII Keys* encrypt the recipe for the *Stone of Phalaja*, which has little to do with archetypal alchemy. In this scenario, the author was a totally original thinker and *The XII Keys* encrypts the recipe for his unique *Phalaja Stone*, yet his overt claim to reveal the *Great Stone of the Ancient Philosophers* would seem completely unfounded and inappropriate to describe a new and unique proprietary product. The question as to "Why the new name?" must also be addressed. The production and use of ethanol would need to be explained in the context of *The XII Keys* tutorial and imagery. In this scenario, *The XII Keys* does not reflect the archetypal process for creating the Philosophers' Stone, but rather an evolved Islamic or European variant described by Valentine as *Phalaja*.

Toward the end of the 4th *Key* however, Valentine makes it perfectly clear that he is describing calcination of stibnite to *flowers of antimony* – a process in antiquity known as *whitening the stone*. The product, *flowers of antimony* and other metal oxides in a generic sense were also collectively known in antiquity by the cover-name *Lime of Stones* (calx from vitriols). The practice dates back to Bronze Age techniques for calcining sulfide ores to their oxide forms as a purification step in the process for creating high-purity bronze. He follows this with an explanation as to why antimony ashes are so crucial to alchemical processes, the understanding being that from *Ashes* we derive *Salt* in a general sense:

> *Workmen prepare Lime of Stones* [vitriols; sulfide ores] *by burning them* [calcination], *that it may be fit for their use ...*
>
> *Everything being burnt to Ashes by Art will yield a Salt*, it in the Anatomizing thereof you are able to keep apart its Sulphur and Mercury and again restow them to their Salt, according to the pure method of Art; then may you again by means of Fire, make thereof again, what it was before its destruction ...
>
> If the Artist [is] want [for] Ashes, he cannot make Salt for our Art, for without Salt our work cannot be made into a body ...
>
> *After that burning* [union], *a new Heaven and a new Earth shall be formed, and the new Man* [Adam; Earth] *shall more gloriously shine forth*, than ever he lived in the old World, for he shall be purified.

The value and importance of Valentine's antimony *Ashes* towards creating his *Salt of Ashes* being understood, the question is by what process is the *Salt of Ashes* brought to perfection?

> *When Ashes and Sand are well maturated and concocted in the fire, then the Artist turneth it into Glass*, which afterward will endure in the fire, and in *colour like a transparent Stone, and is not any more like Ashes*; and this to the ignorant is a great Mistery ...
>
> *Ashes* = antimony ash / flowers of antimony ($Sb_2O_3 \cdot Sb_2S_3$)
> *Sand* = gold nanoparticles / dragon's blood (Au)
> *Glass* = gold-antimony glass / salt of the philosophers' ($Au \cdot Sb_2O_2SO_4$)

The passage above states that a *Glass* is created from *Ashes* and *Sand*. While this makes perfect sense as regards glassmaking, the process he is describing is an allegorical reality meant to conceal the fact that the *Glass* here was *gold-antimony glass*, common to archetypal alchemy and the foundational beginning material for much of his acetate tincture alchemy. Creating a *gold-antimony glass* (Au·$Sb_2O_2SO_4$) can be bypassed as a shortcut by grinding gold (Au) and antimony oxysulfate ($Sb_2O_2SO_4$) together to serve as a *Salt of Ashes* analog.

The chemical identity for *glass of antimony* is antimony oxysulfate ($Sb_2O_2SO_4$). Valentine actually synthesizes this in Triumphal Chariot of Antimony (pages 113 and 115). He does this by creating antimony oxychloride first, reacting it with hot sulfuric acid then cleaning it with ethanol, which can be written as follows:

> *Take one part, of finely pulverized Antimony, and "pulverized salt-armoniac,* so called because it comes from Armenia ; mix these together, place in a retort, and distil together. ...*

$$2\ Sb_2O_3 + 12\ NH_4Cl = 4\ SbCl_3 + 3\ O_2 + 12\ NH_4$$

> *On the product of this distillation pour hot, distilled (common) rain water, removing thereby every salt and acrid taste. Then the Antimony will be of a pure, brilliant, and feathery white;*

$$SbCl_3 + H_2O = SbOCl + 2\ HCl$$

> *dry with subtle heat, place in the circulatory vessel called pelican, pour to it highly rectified spirit of vitriol, and circulate till they are properly amalgamated.*

$$2\ SbOCl + H_2SO_4 = Sb_2O_2SO_4 + 2\ HCl$$

It is plausible, if Valentine's instructions are taken literally, to interpret this as a process to create synthetic glass of antimony in a chemically pure form. This can be brought to a molten state to replicate a more traditional "glass".

KEY IV – SALT OF ASHES

Although it is not expressly stated, Valentine may have created his *Salt of the Philosophers* by fusing *Ashes* (flowers of antimony) with *Sand* (gold) to create a *Glass* (gold-antimony glass). He likens this to turning the inside (seeds of gold and antimony) outwards via initial purification, and then turning its outside inwards via fusion/homogenization. The desired product is an incombustible flux while in fusion that cools to a *Glass*, or as described above, *like a transparent Stone*:

> *Salt out of Ashes is of very great use*, much virtue is contained in them, yet is that Salt unprofitable, unless its inside be turned outwards, and its outside inwards … If you know how to obtain that, *then have you the Salt of the Philosophers*, and the true incombustible Oyl …
>
> Although that many wise,
> Have sought for me with care,
> Yet few consider what,
> My hidden treasure are.

This initial *Salt of the Philosophers*, at once a *Glass* that appeared as a transparent *Stone*, would later come to be known as the *Stone of the 1st Order*. The "hidden treasure" alluded to in the poem above is the gold content combined with, or embedded into the *glass*. This compound was known to Alexandrian alchemists by the cover-names *red-gum*, *latten*, *body* or *earth*. The cover-terms *body* or *earth* are biblical references to the body of Adam, a name that literally means *earth* in early Mesopotamian languages such as Sumerian as well as in Hebrew – the innocent *Adamic-Earth* "corrupted" by gold, destroyed and revivified.

Basil Valentine's 4th *Key* is a rather straightforward process for *whitening the stone* by calcination, creating glass of antimony and uniting it with finely divided gold. According to Valentine, it is a union of *Sulfur* and *Salt* principles, or put another way, a union of *Body* and *Soul* still awaiting *Mercurial Spirit* to animate it. Basil Valentine's *Salt of the Philosophers* is the first of two alchemical compounds required to confect the

BOOK 1 – PREPARING THE TRIA–PRIMA

Philosophers' Stone. During the Alexandrian period, this compound was known by the cover-terms *Red-Man*, *Red-Gum*, *Latten*, *Body*, *Sol*, *Father*, *Old Man*. Islamic alchemists understood this compound as *Kibrīt* (كبريت) meaning *(Philosophers') Sulfur* that found union with *Philosophers' Mercury* detailed in the upcoming 5th *Key*. European alchemists knew it as the *Sophic Gold Calx*, *Stone of the 1st Order*, the *Green-Lion Devouring the Sun*, or simply *Lion's Blood*. Throughout this Key, Valentine employed the following cover-terms in reference to this compound:

- Adam
- Earth
- New Man
- Lion's Blood
- Salt of Glory
- Salt of Ashes
- Salt of the Philosophers
- Double Fiery Man

The above-said cover-terms fully accord with archetypal alchemy, as demonstrated in *Cracking the Philosophers' Stone*, and simply do not lend themselves to interpretation as potassium salt in any reasonable way. Confecting the Stone was often viewed by Christian alchemists such as Valentine and others as modeling the creation and birth of Adam.

The chemical identity of the *Earth* (mundi; ☿) by which Adam's body-soul was created (double fiery man; new man; ☿), is purified and reduced *gold-antimony oxysulfate* ($Au \cdot Sb_2O_2SO_4$) in the form of a dark reddish-brown powder, which is then infused with soul-spirit present in the product of the next Key.

In Valentine's Philosophers' Stone model, *Body-Soul* is *gold-antimony glass* ☿, whereas *Spirit* ☿ is a distillate in the form of the liquid salt-menstruum employed in the synthesis.

Gold	**Antimony**	**Divine Water**
Soul	Body	Spirit
(immortality / incorruptibility)	(growth)	(regeneration)

KEY IV – SALT OF ASHES

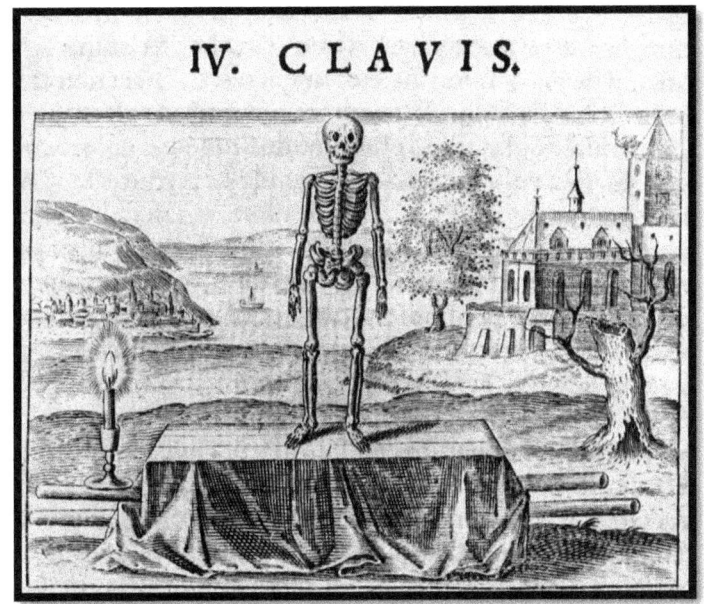

Symbolism of IV Clavis

- **Skeleton** = "ashes to ashes, dust to dust", death, ashes, body contains no life
- **Candle** = light of knowledge of the Salt of the Philosophers and how to proceed
- **Dead tree** = dead and hollow, just as the Salt of the Philosophers until it becomes the Stone
- **King's palace** = the product of the 5th Key to come, secret solvent, nearer than in the 1st and 3rd Keys
- **Bird of Hermes / Eagle** = looking at the sea from the tower, suggestive of the product of the 5th Key to come
- **Living tree** = dead tree in foreground and living tree in background suggests new life to come
- **City by the sea** = the King's kingdom relies on the salty sea for its sustenance
- **Salty sea in the distance** = in earlier Keys, a salty-sea indicated aqua-regia and this may indicate that its use is finished, or behind the body, but with sail-boats and its city also suggests activity and life

Key V – Spirit of Mercury

The Philosophers Mercury, and not the Vulgar, being reduced unto water, dissolveth the Philosophick Salt together with the Purple Mantle ... for it is Mercurius Duplicatus.

Basil Valentine

Basil Valentine's *5th Key* encrypts the recipe for an inorganic nonaqueous liquid antimony compound that he referred to as the *Spirit of Life* or *Spirit of Mercury*. His *Spirit* is depicted as a vapor, breath or woman associated with flowers in general, or specifically with the *Rose of Sharon*, *Lily of the Valley* or red *Opium Poppy* in earlier traditions. At this stage in the recipe for confecting the Philosophers' Stone, the process of creating the two chemical compounds comes to completion – the second and final being the *Spirit of Life* addressed in this Key. Valentine alludes to *Salt of Philosophers* (*Soul of the Earth* below) from the previous Key as requiring the *Spirit of Life* to enter into and enliven it.

> *For that which is dead cannot adde to that which hath life, and the off-spring of the dead cease, because the Spirit of Life is wanting; therefore the Spirit of Life and Soul of the Earth, that dwelleth in it, and operateth on Earthly things, from the Cœlestial and Syderial. ... the Earth affordeth not those virtues of its self, but the living Spirit which is in it; and if the Earth should be without that Spirit, it were dead, and could not yield any more nourishment ...*

> *For the Spirit is the Life, which is nourished by the Stars ... as the Mother preserveth the Fetus in the Womb, and seedeth it there; so also doth the Earth nourish in its Bosom the Minerals, by its Spirit received from above.*

BOOK 1 – PREPARING THE TRIA-PRIMA

> **Cœlestial** = *heavenly* or *divine* to indicate antimony ($Sb_2O_3 \cdot Sb_2S_3$)
> **Syderial** = *of a distant star* to indicate sal ammoniac (NH_4Cl)
> **(Soul of the) Earth** = *dry compound* of the 4th *Key* ($Au \cdot Sb_2O_2SO_4$)
> **Spirit (of Life)** = *liquid distillate* of the 5th *Key* ($SbCl_3$), nourished (created) by *Stars* (NH_4Cl; ✶)

According to Valentine, the earthy compound of the 4th *Key* will become impregnated with liquid *Spirit*, its offspring and chemical identity being the product of this Key and depicted in alchemical imagery as a child. He sees this as an analogy to the composition of our own being:

> *When* a Man looks in[to] a Glass [mirror], *there is the reflection of his Image, which if you go to touch with your hands, you find nothing tangible but* the Glass *wherein the person looked: So* also from this matter must be drawn a visible Spirit, *which nevertheless is impalpable. That very same Spirit, say, is the Radix [or root; origins] of the Life of our Bodies, and the Mercury of the Philosophers, from whence our liquid Water is prepared in our Art … For* our beginning is a secret and palpable body *[Soul of the Earth],* the middle is a fugitive Spirit *[Spirit of Life], and a golden water without any corrosive …*

In the true double entendre style of literary expression common to most alchemists, he also encrypted the cover-terms of specific alchemical substances. In the passage above, *glass* = *glass of antimony*, which for Valentine is the *prime material* for many of his alchemical creations. His solvent is indeed a distilled *spirit*, but it does not necessarily appear like, nor function like other mineral acids – it does dissolve metals including silver and gold, yet also coagulates them to a salt.

In the passage below, he summarizes the process thus far, beginning with gold's purification and particle-size reduction, the creation of a white compound (5th *Key*) and a red compound (4th *Key*), further indicating that the gold, glass and solvent once combined and heated, result in the Philosophers' Stone.

KEY V – SPIRIT OF MERCURY

> *For a Co[n]clusion of these things, I tell you for a truth, that one work proceedeth from another; for our matter must be very well and highly purified in the beginning of our work, then dissolved and destroyed, and th[o]roughly broken and reduced to dust and ashes [Cœlestial Sulphur; Au]: When this is all done, then make thereof a volatile Spirit white as Snow [Spirit of Mercury; SbCl₃], and another volatile Spirit red as Blood [Salt of Ashes; Au· Sb₂O₂SO₄], which two Spirits contain in them a third, and yet are but one Spirit. These are the three Spirits, that preserve and prolong Life, joyn them together ... keep them in a warm bed until the perfect time of their Nativity; then shall you see and understand what the Creator and Nature hath discovered unto you; and know that my lips never yet so plainly revealed anything.*

Valentine concludes with the declaration that he has spoken plainly and indeed, he has if the aspirant can interpret his cover-terms correctly. The trinity he is describing is addressed elsewhere and provides a solid foundation by which to understand Valentine's template for creating the archetypal Philosophers' Stone:

> *Hermes saith, these things are required for the work, first a volatile, or Mercurial water, aqua celestis [SbCl₃], the Leo viridis, which is the Philosophick Lion [Sb₂O₂SO₄], thirdly aes Hermetis, Sol or Ferment [Au].*
> – The First Treatise of the Sulphur, vitriol and Magnet
> of the Philosophers

Mercurial Water / Aqua Celestis = butter of antimony; $SbCl_3$
Leo Viridis / Philosophick Lion = green glass of antimony; $Sb_2O_2SO_4$
Aes Hermetis / Sol / Ferment = finely-divided gold powder; Au

Supporting Documents

The above is also stated in a similar manner in *An Elucidation of the XII Keys*:

> *In th[e]se, together driven goldish waters lieth hid that true bird and Eagle [SbCl₃], the King with his heavenly Splendor [Au] together with its clarified Salt [Sb₂O₂SO₄], which three you find shut up in this one thing*

BOOK 1 – PREPARING THE TRIA–PRIMA

and golden property, and from thence you will get all that, which you have need of for your intention.

True Bird [of Hermes] and Eagle = antimony ash + sal ammoniac
King with his Heavenly Splendor = pure, fine gold powder
Clarified Salt = glass of antimony

... the *Mercurial Spirit is cold and moist*, the *sulphureous Soul is warm and dry*, and *this liquor is the true prima material*, and first seed of Metals and Minerals ...

Cold and Moist + Warm and Dry = *Tetrasomia* as a *Union of Opposites*

In my former writings, as also *in the XII Keys*, from the first to the last, I ordered thus my stile of writing, wherein I held forth unto posterity the practick, *how the great stone of Philosophers, or the best purified gold, may be made out of Sulphur and Salt, with the help of the spirit of Mercury*, which must be *drawn from a crude unmelted Minera, according to the Tenor of my fifth Key* set down in a parabolical manner.

The symbolic double entendre for *Spirit of Mercury* encrypted in Valentine's celebrated *2nd Key* also appears in written form in the supporting text, *Appendix And plain Repetition or Reiteration of Basil Valentine*, where the substances and processes can be interpreted as *aqua regia* and / or *butter of antimony*:

The *spirit of common Salt*, which is drawn after a peculiar manner, maketh Gold and Silver volatile, if a small quantity of the *spirit of the Dragon be added* to it, it dissolveth it, and carrieth it over with it per Alembicum, *as also doth the Eagle with the Dragons spirit* ...

The chemistry in the above passage remains accurate in both cases. *Spirit of common Salt* is hydrochloric acid and the *Eagle* is ammonium chloride. Either of these, when reacted with *Dragon's Spirit* as potassium nitrate or nitric acid, creates a form of *aqua regia*. Likewise, when *Dragon's Spirit* = *flowers* or *glass of antimony*, they each result in crude antimony trichloride. It was a fantastic use of wordplay to encrypt the identity of *Philosophers' Mercury*, the most prized and important nonaqueous mineral solvent in archetypal forms of alchemy, yet disguised as a lesser solvent.

Butter of antimony could be reacted with ethanol to achieve a medicinal *sweet Oyl* as alluded to in the *Appendix And plain Repetition or Reiteration of Basil Valentinus*:

> *For a Conclusion of this Appendix, I must needs tell you that out of black Saturn [stibnite] and fri[e]ndly Jove [corrosive sublimate], a Spirit may be extracted, which is afterwards reduced into a sweet Oyl, as its noblest part, which Medicine, particulariter doth most absolutely take[n] away the nimble running quality from common Mercury, and bringeth him to a melioration, as I taught you before.*

In the above passage, Valentine is creating antimony trichloride via a distillation of stibnite and mercuric chloride, resulting in mercuric sulfide byproduct known to alchemists as *cinnabar of antimony*, described poetically by Valentine as having *take[n] away the nimble running quality from common Mercury*. The unique medicinal *sweet Oyl* alluded to is created from *butter of antimony* and ethanol, and was presented in precise detail in the *Triumphal Chariot of Antimony*. The primary difference as regards distilling *butter of antimony* in the above passage and the version in the *Triumphal Chariot* is that in the latter, Valentine clearly gives the recipe for *butter of antimony* created with sal ammoniac:

> *Take one part of finely pulverized Antimony, and pulverized salt-armoniac [equal quantities] … mix these together, place in a retort, and distil together.*

He goes on to describe the product's conversion to *Algarot Powder* and from there a *sweet Oyl.* Ethanol and acetate-tincture oils however, are contextually different from the *Oil* as butter of animony mentioned in the *Appendix and plain Repetition or Reiteration*:

BOOK 1 – PREPARING THE TRIA–PRIMA

> First know, that no common Argent vive is fit for our use; but our Argent vive is made of the best Metal by the Spagirick Art, pure, subtle, clear, splendent, as a Fountain, transparent as Christial [Crystal], without any impurity; of this make a Water or incombustible Oil …

A passage found in *The First Treatise of the Sulphur, Vitriol, and Magnet; Section III Of the Philosophers Magnet*, reinforces the dissolving and coagulating properties of Philosophers' Mercury or *Magnesia*, which was viewed as a type of *magnet* holding gold and antimony in perfect conjunction – ratios for which are the topic of the 6th Key.

> Thus the coagulated Mercury must by Art be turn'd into its tria material, or water, that is, Mercurial water. This is a stone and no stone, of which is made a volatile fire, in form of a water, which drowneth and dissolveth its fix't father [gold] and its volatile mother [flowers of antimony]. …

During the Alexandrian period, this compound was known by the cover-terms White-Wife, White-Gum, Eudica, Body, Luna, Mother, and Old Woman among others. Islamic alchemists understood this compound as zībaq (زئبقي) meaning [Philosophers'] Mercury, which united with Philosophers' Sulfur detailed in the previous 4th Key. European alchemists knew this liquid solvent as Sophic Mercury, Azoth, Ignis Aqua, Our Vinegre, Sword, Magnet or simply Eagle's Gluten. Throughout the 4th Key, Valentine uses the following cover-terms in reference to this compound:

- King's Palace
- Mercurial Spirit
- Mercurial Water
- Mercurial Oil
- Argent Vive
- Starry Water
- Aqua Cœlestis
- Spiritual Water
- Salty Sea
- Aqua Permanens
- Magnet
- True Bird
- [of Hermes]
- Eagle

In the sulfur-mercury theory, it served as the mercury principle, whereas in the Tria Prima (three primes; salt-sulfur-mercury) theory of Al-Razi and

KEY V – SPIRIT OF MERCURY

Paracelsian Spagyrists and Iatrochemists it was the combined *Mercury* principle, viewed as the *Spirit* of the Philosophers' Stone. Valentine drew strong parallels to infusing otherwise dead matter with living spirit:

> *To sum up all, all things, nothing excepted, that may be handled and felt are Natural, but they must be made Supernatural … For the supernatural alone hath in it a lively and quick virtue to work, but the Natural hath but a dead palpable form.*

Valentine was again speaking in double entendre when he wrote the above passage. It is rich with spiritual or religious overtones, yet served equally well for material substances and processes.

Unlike the directness of the 4th *Key*, Basil Valentine's 5th *Key* abounds in classical mythology, symbolism and layers of hidden meaning. The woodblock print indicates Valentine's familiarity with the *Metamorphoses* (The Golden Ass) of Apuleius, in which the plot revolves around Lucius' curiosity and insatiable desire to see and practice magic.

In the tale of Cupid and Psyche, the fructifying breath/wind of Zephyr transports Psyche to Cupid (Eros) and an eagle gathers water for her – all very alchemical themes. Incredibly, the product of this Key was created by Valentine by at least three recorded recipes. It was viewed as the animating spirit that brought the dead body (skeleton of the 4th *Key*) back to life. It later became known as *butter of antimony*, which can be purchased easily from a chemistry supply as a shortcut to the 5th *Key*. Basil Valentine's *Spirit of Mercury* is the second of two alchemical compounds required to confect the Philosophers' Stone. The chemical marriage of these is addressed next in Valentine's 6th *Key*.

KEY V – SPIRIT OF MERCURY

Symbolism of V Clavis

- **Sun** = Sol, gold (also the black sun, Kronos-Saturn-Shamash in mythology)
- **Lion** = Lion's blood, King (product of 4th Key) united to the Queen (product of 5th Key)
- **Cupid** = Blind love, a sharp-tipped golden arrow indicates an eternal love-union of Eros and Psyche
- **Psyche** = (Greek Ψυχή, "Soul" or "Breath of Life") daughter of the King and Queen, wife of Cupid (Eros)
- **Zephyr** = (Ζέφυρος, "the west wind") abducted Chloris of the flowers who gave birth to Carpus (fruit)
- **Breath of Zephyr** = to breathe life into matter, to enliven via distillation or chemical union
- **Retort** = belly of Hermes, "and the wind carried it in its belly" to indicate distilling Hermes / Mercury
- **Flowers** (of antimony) = prime material for creating the Red King (4th Key) and White Queen (5th Key)
- **7 Flowers** = 7 planetary stages symbolized by an extended color-regimen
- **Bellows** = wind, breath (means soul or spirit in many languages) indicates distillation / revivification

BOOK 2

Confecting the Stone of the Ancients

*Dear Friend and lover of Art,
In my Preface I promised to shew unto thee ...
That Corner Stone, and that Rock,
so far as I am permitted from above,
as our Ancestors the Ancients prepared their Stone,
which they attained from the Most High,
for the preservation of their health,
and for the benefit in this present world.*

Basil Valentine

Key VI – Chemical Wedding

... the fiery King will exceedingly love the pleasant voice of the Queen, and out of his great love embraceth her, and satiateth himself with her, until both vanish and become one body.

Basil Valentine

Basil Valentine's 6th *Key* encrypts the precise measure and ratios of *Earth* and *Water*, including the volume of *Air* necessary in the digestion vessel. Overtones of love and sexual union, seeds, fertile growth and reproduction are worked into a highly dualistic theme of balance between male and female principles – complete and perfected upon conjugal union:

> *Man without a Woman is esteemed but as half a body, and a Woman without a Man likewise obtaineth the name but of half a body, for either of them by themselves can produce no fruit; but when they live together in a Conjugal State, the body is perfect, and by their Seed an increase succeedeth.*

Valentine quickly transitions to cautionary examples of improper measures in an apparent effort to reinforce the notion that, like Natural processes, harmonious balance between his alchemical compounds is necessary for success:

> *When too much seed is cast on the ground ... mature fruit cannot be expected; and if there be too little seed ... no profit can be expected.*

> *... give his Neighbor just measures, and let him use just weights and measures, then he avoideth curses, and gaineth the blessings of the poor.*

> *In great waters it is easie to be drowned, and shallow waters are easily exhausted by the heat of the Sun, that they are of no use.*

In the passages above, Valentine is clearly warning the reader that the measure of *Water* must be just right. His instructions to *let him use just weights and measures* can be interpreted as hinting that Valentine's ratios are based on mass rather than volume, yet he clearly addresses volume later in this Key as well:

> *Therefore to obtain your desired end, a certain measure must be observed in the commixtion of the Philosophick Liquid Substance, that the greater part do not over-power and over-press the lesser, whereby the Effect will be hindered, and lest the lesser be too weak for the greater, but let there be made an equal dominion ... if Neptune hath rightly prepared his Water-bath, then take a just quantity of the Aqua Permanens, and have a great care that you take not too much, nor too little.*

After making the obligatory warnings regarding ratios, he begins to address volumetric requirements of the glass digestion vessel, which he encrypts in the cover-name *World*. He explains in *An Elucidation of the XII Keys* that the powdered *Earth*, referred to below as *a double fiery man*, is to take up 25% of the volume of the vessel:

> *A double fiery man must be fed with a white Swan which will kill each other, and will again revive. And the Air of the four parts of the World must possess three parts of the included fiery Man, that the song of the Swan may be heard, when she harmoniously sings her farewell, then the ro[a]sted Swan will be food for the King, and the fiery King will exceedingly love the pleasant voice of the Queen, and out of his great love embraceth her, and satiateth himself with her, until both vanish and become one body.*

The above passage essentially instructs the aspirant to fill the digestion vessel to ¼ its volume with *double fiery man* – interpreted here as finely powdered *gold-antimony glass*. The passage that follows redirects the aspirant to the *6th Key* for the method of conjunction, and then hints that *Keys 7-10* result in the Philosophers' Stone:

> Then proceed unto the practick and conjunction, *and have a care, that you be provident therein, that at their conjunction you* do not too much to the one, nor too little to the other, take notice of the quantity, *and observe exactly the division of the seeds, hereunto* minister a certain measure, and mark my sixth Key, *then* proceed in the begun process, according to the order of the seventh, eighth, ninth, and tenth Key's ... go on with it *to the appearance of the Kings honour and glory, to his highest purple garment* ...

Valentine appears to suggest an equal measure by weight of the two compounds. He likens their union to the following:

- Love or embrace between a King and Queen
- Sexual union between a Man and Woman
- A fiery [red] Man/King dining on white Swan
- Seed and adequate Water to produce Fruit

He concludes the chapter by alluding to the fact that everything up to this point concludes the Magistery, which is to say informed knowledge of the sacred *Art* of confecting the Philosophers' Stone.

The *Rock* mentioned in the following passage is stibnite brought through all of its manifestations or expressions into union with gold and thus united perfection as the Stone. Perfection according to Valentine means a balance of the four elements – Tetrasomia (earth, water, air, and fire) brought into harmonious equilibrium. It is the alchemical "world in a glass" in which the *World* or *Artificial Heaven* is the reaction vessel with its chemical elements undergoing growth and transformation in an evolutionary process towards perfection. Valentine's *Earth* is a mound of powdered *Stone* (i.e. Stone of the 1st Order; gold-antimony glass) at the

BOOK 2 – CONFECTING THE STONE OF THE ANCIENTS

bottom of the digestion vessel, to which *Water* (i.e. divine water; molten *butter of antimony*) descends from above in perfect proportion, the atmosphere hermetically sealed and the whole warmed by an external *Fire* that catalyzes the internal *Fire* – this being chemical reactivity. It is a world in miniature, the alchemical microcosm, with all of its elements in allostatic movement towards perfect homeostasis.

> *The Knowledge of our Magistery is herein very necessary for division, and conjunction must be rightly made, if Art is to produce riches, and the Scales must not be falsified by unequal weights. This is the Rock we proposed, that you be sure to finish this Work by an artificial Heaven, by Air, and by Earth, with true Water and perceptible Fire, in giving of a lawful weight without any defect, as I have rightly informed you.*

Valentine's use of the cover-name *Rock* for antimony is an oblique reference to two biblical passages that address the setting-stone of the Temple of Solomon:

> *I Chronicles 29:2* So *I have provided for the house of my God, so far as I was able, the gold for the things of gold, the silver for the things of silver, and the bronze for the things of bronze, the iron for the things of iron, and wood for the things of wood, besides great quantities of onyx and stones for setting, antimony, colored stones, all sorts of precious stones and marble.*

> *Isaiah 54:11* O *afflicted one, storm-tossed and not comforted, behold, I will set your stones in antimony, and lay your foundations with sapphires.*

For Valentine, comprehension of the process thus far, of each Key up to the 6th, meant that the Philosopher was in possession of the Stone. The rest was merely a matter of attending to the temperature and varying its degree in response to visual cues and color-indicators observed inside the vessel.

Plainly stated, Basil Valentine's 6th *Key* encrypts instructions for filling a glass digestion vessel to 25% of its capacity with powdered gold-antimony glass (Au·Sb$_2$O$_2$SO$_4$), and adding liquid antimony trichloride (SbCl$_3$) to it from above drop-by-drop (symbolized by rain) in approximately an equal proportion by weight. Two fires are hinted at, these being 1) external fire that controls the heat regimen to follow, and 2) an innate fire understood today as the potential for chemical reaction. During the period in which Valentine wrote and afterward, the process described above was encrypted by the cover-term *Chemical Wedding*. From this point onward, success at confecting the Philosophers' Stone relied on knowledge of the temperature regimen addressed next in Valentine's *7th Key*.

BOOK 2 – CONFECTING THE STONE OF THE ANCIENTS

KEY VI – CHEMICAL WEDDING

Symbolism of VI Clavis

- **King** = male compound ($Au \cdot Sb_2O_2SO_4$)
- **Queen** = female compound ($SbCl_3$)
- **Priest** = uniting principle, the alchemist, God, wisdom, etc.
- **Neptune** = preparing his water bath, a just quantity (correct measure) of aqua Permanens
- **Alembic** = measure of Air in vessel (75% of the volume)
- **Trident** = "goldish waters that hide the three" (tria prima)
- **Furnace** = 2 fires (external and innate) dual breaths of life from the belly of the furnace / in the vessel
- **Atmosphere** = Sun (gold), Moon (antimony), Rain (flux) as the chemical wedding
- **Swan/Bird of Hermes** = the King's first roasted meal, "Ancients have compared it with a volatile Bird"
- **Rainbow** = three color-indicators of the synthesis (black, white, red)

Key VII – Judged by Fire

If we rightly behold our own souls, then shall we be made Sons and Heirs of God ... But this cannot be done, unless the Waters be dryed up, and Heaven and Earth with all Men be Judged by Fire.

Basil Valentine

Basil Valentine's *7th Key* encrypts the heating regimen to confect the Philosophers' Stone. The chapter begins with a discussion of the nature and importance of natural heat or fire:

> *Natrual heat preserveth the Life of Man, for if that be gone, Life Ceaseth. Natural fire, if it be moderately used, defended against cold, but too much is destructive.*

Valentine then segues into a lecture on the ramped heating regimen and finishes with instructions for filling the vessel, adding the reagents and sealing it airtight. He presents his lessons in an order that emphasizes knowledge of temperature control first, followed by practical instructions for handling the reagents and glassware. His purpose in the order may have indicated order of importance or alternately as a diversion mechanism aimed at qualifying the would-be adept, and / or distracting the less-worthy. Since temperature control comes after the experiment set-up, the thematic order has been adjusted in this treatment towards a more logical flow.

Salt, Water and the Seal of Hermes

The previous chapter ended with instructions for the proper measure of *Earth*, a.k.a. *Salt of Glory* (the dry reagent) and *divine water* (the liquid reagent), called *Spiritual Water* in the passage below. *Salt* was placed in the bottom of the digestion vessel occupying 25% of the volume and an equal weight of *Spiritual Water* was added to it drop-wise (Latin: *guttatim*). The following passage picks up from where the previous chapter left off.

> ... *take your Spiritual Water, whereon the Spirit moved at the beginning, and shut the door* of defence upon it; for from that time shall the *Heavenly City be besieged by Earthly Enemies,* and your *Heaven must be strongly defended with three fences and walls,* that there be no entrance but one, and let that be very well guarded.

Heavenly City = reaction vessel
Heaven = atmosphere inside the reaction vessel
Earthly Enemies = corrosive vapors, hydrogen chloride
3 Fences and walls = hermetic seal (1-stopper, 2-twine or wire, and 3-sealing wax)

Alchemical Chaos

When the digestion vessel is filled and sealed, the experiment is likened to the original *chaos* from which matter came into being. *Chaos* was understood in antiquity as the formless primordial state just prior to the separation of *heaven* and *earth* during the creation of the universe or cosmos. In the alchemical sense, it refers to the gaseous vapors that will begin to emanate from the composition at the bottom of the vessel, which can be described quite literally as a separation of *heaven* (acidic atmosphere) and *earth* (semi-solid matter) upon being gently heated. Gentle heating is called for initially, which releases hydrogen chloride gas at a very low temperature as antimony trichloride decomposes in the vessel.

> When all these things are done, *kindle your Philosophick Lamp,* and seek what you have lost, *give so much light as may suffice ...*

KEY VII – JUDGED BY FIRE

Winter – Sophic Decomposition

The heating regimen is likened to the four seasons of a temperate climate, beginning with winter. *Sophic decomposition* is catalyzed at 40-42 °C. The parallel to winter is appropriate to this stage. Just as all plant-life appears completely dead in winter, so does the matter in the vessel. It will give the impression as though nothing at all is happening for quite some time, as if the matter and its inner fire is either dead or lying dormant in a state of hibernation. Eventually the matter will decompose, turn black and vapors will become visible at the cooler top portion of the vessel.

> *In the Winter the common people count all things dead*, because *the cold bindeth the Earth* that nothing can grow ...

Spring – Vegetative Germination

A transition stage, known traditionally as *vegetative germination*, is catalyzed at 47-50 °C. The matter takes on a dark greenish coloration spreading to yellow before transitioning to grey, described by Valentine as *fair, amiable and various colours*.

> ... but as soon as the *Spring appeareth*, that the cold lesseneth *by the ascent of the Sun, all things revive*, Trees and Herbs grow, and Insects which hid themselves from the cold Winter creep forth out of their holes and caves in the Earth; all Vegetables yield a new savour, and their Excellency is discovered by their *fair, amiable, and various colours* of their Blossoms ...

Summer – Volatilization through Liquid

A liquefying stage, known traditionally as *volatilization through liquid (evaporation)*, begins like a frost or snow that spreads throughout the matter as it transitions from grey to white catalyzed at 70-80 °C or higher. The matter will typically liquefy to a whitish solution at this point so long as the appropriate ratios have been used.

> ... *Summer continueth the operation, and bringeth forth Fruits from these several kinds of Flowers: For which thanks be given to the Creator, who by his Ordinance hath set bounds unto these things by Nature.* ...

> *When the Sun declineth from us in the Winter, it cannot dissolve the Snowy Mountains, but when it approacheth nearer in the Summer, the Air is hotter, and more powerful to dissolve the Snow, that it turneth it into water, and destroyeth it: For the weak must yield to the stronger, and the stronger over-ruleth the weak.*

Autumn – Volatilization through Dry Earth

The drying stage, known traditionally as *volatilization through dry earth (sublimation)*, is catalyzed at 100-110 °C. At this stage, the matter begins to crystalize to an amorphous powder.

> *Thus also in our Magistery the government of the Fire must be observed, that the moist Liquor be not too suddenly dryed up, and the Philosophick Earth too suddenly melted and dissolved* ...

> *The Vine hath more need of the heat and beams of the Sun at the latter time of its maturation, than it had in the beginning of the Spring: And if the Sun doth strongly operate in the Autumn, the Vine doth yield a better and stronger Juice, than if the heat of the Sun-beams be weak or deficient.*

Themis / Iustitia – Goddess of Justice

The notion of accurate discernment or discrimination is a strong theme in Valentine's 7th *Key*. The alchemists' ability to accurately judge conditions in the vessel by various visual indicators is ultimately crucial to success. Valentine symbolized this important fact by the image of Themis-Iustitia, the Greco-Roman goddess of justice; most often depicted blindfolded holding a set of scales in one hand and a double-edged sword in the other. This imagery is very ancient, dating back to the ancient Egyptian personification of truth, balance and order as the goddess Maat, and later as Isis.

KEY VII – JUDGED BY FIRE

The Goddess of Justice is an appropriate theme for Valentine's 7th Key for the following reasons:

- Themis literally means "that which is put in place", representing divine order or natural law
- She was at Delos to witness the birth of Apollo, God of Enlightenment (Truth, Light, Art, the Sun)
- As a consort of Zeus she controlled time, specifically the rightness of order unfolding in time
- Justice knew no middle ground, for her there was only correct or incorrect

It is not clear whether Valentine intended the image of Themis to indicate the final fixation stage of the heating regimen, or to symbolize good judgment. In either case, it is a very appropriate representation. This stage is rarely discussed in alchemical literature and is understood as fixation, or in alchemical terms *judgment by the fire*, resulting in the substance being considered fireproof. It was typically implemented as a single step but could be divided into two. Following crystallization, fixation occurs as the product melts with further heating between 120-130 °C. At increased heat, the matter decomposes releasing free chlorine vapors. The process can be continued to around 170 °C or even much higher. Upon cooling and drying, the matter appears as a heavy reddish-purple to brown colored amorphous powder and was considered fireproof, or *judged by Fire*.

> *If we rightly behold our own souls, then shall we be made Sons and Heirs of God, to effect that which seemeth now impossible to us: But this cannot be done, unless the Waters be dryed up, and Heaven and Earth with all Men be judged by Fire.*

Basil Valentine's 7th Key encrypts the instructions for preparing the reagents in the vessel, creating an airtight seal and informed knowledge of the temperature regimen for all stages of confection. The information represented in the 7th Key is fundamental towards success at achieving the finished Philosophers' Stone. Correct application of the heating regimen however, requires skill of observation and discernment symbolized in the accompanying woodblock print by the figure of Justice without the customary blindfold. As is usually the case with alchemical imagery, Justice is represented in a gender-neutral or non-specific manner. From this point onward, the aspiring adept must be able to recognize specific visual indicators as they occur and adjust the temperature accordingly. The initial indicators that signal success arrive in the form of a vapor at the top of the vessel along with liquefying and decomposition known in alchemical terms as putrefaction, the subject addressed next in Valentine's *8th Key*.

KEY VII – JUDGED BY FIRE

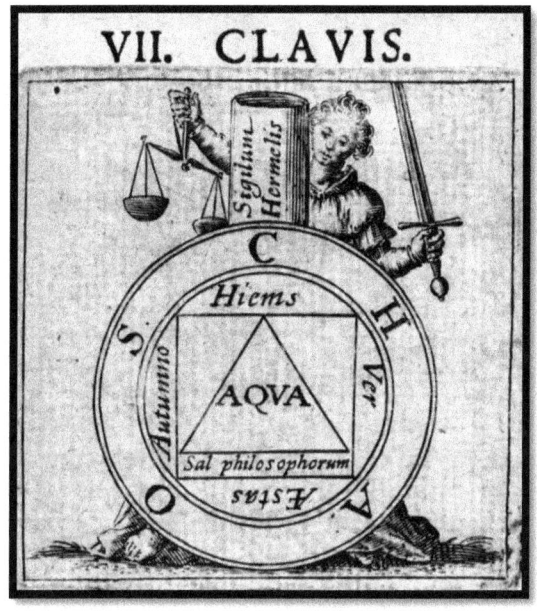

Symbolism of VII Clavis

- **Sigilum Hermetis** = Seal of Hermes, airtight seal of the vessel
- [Ω] **C H A O S** = the All, formless being / infinite empty space / the lower world (microcosm)
- [□] **Sal philosophorum** = Salt of the Philosophers at the bottom of the vessel
- [△] **Aqva** = [Divine or fiery] Water added to salt from above
- **Hiems** = Winter / decomposition (black stage at 40-42°C)
- **Ver** = Spring / vegetative germination (transition stage at 47-50 °C)
- **Æstas** = Summer / volatilization via liquid (white stage 70-80 °C)
- **Autumno** = Autumn / volatilization via dry earth (red stage 100-110 °C)
- **Themis/Iustitia** = Judgment by fire / fixation (melt and recrystallize 120-130 °C or higher)
- **Themis un-blindfolded** = discernment or judgment based upon visual observation
- **Scales and sword** = knowledge of ratio or measure of reagents and volumetric space requirements
- **Circular symbol in front of Themis** = sealed long-neck round-bottomed glass reaction / boiling vessel

Key VIII – Putrefaction of Adam

So Adam was first brought forth, generated and compounded of Earth, Water, Air, and Fire: of Soul, Spirit, and body; and of Mercury, Sulphur, and Salt.

Basil Valentine

Basil Valentine's 8th Key encrypts knowledge of the primary visual indicator that the synthesis will be a success. For Valentine and a great many other alchemists, putrefaction was the long awaited sign that every stage thus far in the process of confecting the stone had been performed correctly. The experiment as summed up in the 7th Key was subjected to low heat for quite a lengthy waiting period. Eventually the matter begins to decompose, taking on the color and consistency of black pitch likened to the putrefaction of a rotting corpse, and typically accompanied by visible vapor condensation in the upper part of the vessel. To Valentine, all things and specifically seeds first putrefied before the sprouting of new, or to religious alchemists, resurrected life could occur:

> *All Flesh be it Mans or Beasts yeildeth no increase or propagation, unless it be first putrefied*, also the Seed when it is Sown, and all that is under or belonging to Vegetables *cannot increase but by putrefaction*.

The 8th Key is one of Valentine's more lengthy chapters primarily to elucidate that occult knowledge of putrefaction is one of *Nature's Mysteries* because, he states; *in truth all life proceedeth from and is caused by putrefaction*. The initial putrefaction, called *melanosis / nigredo* (or blackening) signified not only chemical decomposition, but also a religious purging of evil, the *dark night of the soul* that must occur

BOOK 2 – CONFECTING THE STONE OF THE ANCIENTS

prior to rebirth in a new and exalted form. He goes on to list examples of putrefaction as it occurs in nature in the form of sprouts from seeds, chicks from eggs, ants from honey, worms from decayed flesh, etc. and further explains that it is a combination of the transformative powers of earthly elements in combination with stellar influences that brings forth new life from apparent decay. To Valentine, the concordance between the four elements of the *Tetrasomia* (earth, water, air and fire) was the hidden power behind growth and transformation. He points the reader to the reaction occurring in the vessel as a visual model or study aid and veritable proof of the process:

> *... I tell you, who earnestly intend the separation of Nature, and to understand the division of the Elements, that in the distillation of the Earth [in the vessel], first the Air [visible vapors] cometh very easily, then after some certain time cometh the Element of Water [condensation], the [hidden or natural] Fire was included in the Air, because both are of a spiritual Essence, and do both wonderfully love each other. The Earth remaineth in the bottom, wherein is the most precious Salt.*

> *... the Water and the Earth may be taken apart ...*

In plain terminology, Valentine is describing the decomposition of the antimony trichloride component as it releases hydrogen chloride vapor. The vapor condenses in the vessel appearing very much like water. He must have known that these vapors are highly corrosive based on his observation that *Fire was included in the Air*. What he describes is reasonably accurate – indeed, it would appear that the *Water* and the *Earth* had been separated. Heat is necessary to catalyze the reaction:

> *For you ought to know, if any thing proceed by putrefaction it must of necessity be after this manner: The Earth by its secret and hidden moisture is reduced into corruption, or a certain destruction, which is the beginning of putrefaction; for without moisture, as is the Element of water, there can be no putrefaction: For if any generation do proceed from putrefaction, it must needs be kindled and produced by the*

KEY VIII – PUTREFACTION OF ADAM

> *property of heat or Element of fire; for without natural heat no production can be made,* and if that production do assume a living breath and motion, that cannot be without Air ...

Valentine then shifts from Natural Philosophy and classical Greek theories of the four elements (Tetrasomia) to Christian mysticism in his depiction of the reaction occurring within the vessel, again likening the process to the creation of the *first Man, Adam*, yet elegantly interlaced with salt-sulfur-mercury theory.

For al-Razi, Valentine and Paracelsian Spagyrists, all inorganic matter and organic life was composed of solid matter, liquid or vapor and heat. This trinity was known as the *Tri-Prima* or *Three Primes*, understood as *Salt, Mercury* and *Sulphur* principles:

> Know that when *Adam the first Man was formed by the great Creator out of a lump of Earth,* there did not as yet appear any perceptible motion of life, until *God breathed a Spirit into him,* then was that lump of Earth endowed with power. *In the Earth was the Salt i.e. the body, the inspired Air was the Mercury, the Spirit. The air by this inspiration did give a genuine and temperate heat, which was Sulphur,* i.e. Fire, then it moved itself, and Adam manifested by this motion, that a living soul was inspired into him ...

For very religious alchemists, the process of confecting the Philosophers' Stone served as a perfect working model, a microcosmic fractal example of the precise act of man's creation where the Alchemist became the creator in control of the elements of creation – in effect "playing God". Confecting the Stone served as an appropriate objectification of cosmological principles inherent in observable reality. The notion of being able to control these forces on a microcosmic scale as a window into the forces of creation verged dangerously towards heresy and for this reason and others, alchemists published under pseudonyms to protect their identities.

BOOK 2 – CONFECTING THE STONE OF THE ANCIENTS

Eve figures into the alchemical model as being derived from Adam, which is to say that after life had been infused into the Adamic-earth, transfiguration occurred yielding a female form visible as the white and potentially fluid product of the next stable stage of the reaction, known as *Leucosis / Albedo* (or Whitening):

> *So Adam was first brought forth, generated and compounded of Earth, Water, Air and Fire; of Soul, Spirit, and body; and of Mercury, Sulphur, and Salt.*

> *After the same manner Eve the first Woman and Mother of us all, pertook of the same composition, being taken from Adam; so Eve was produced and builded from Adam: which note well.*

Following this very logical and pedagogical thread, Valentine then segues into a summary of creating *Fixed Tincture from Antimony* or *Oil from Antimony*, which can only be construed as an odd inclusion to support his case, or as a clever means to throw the unworthy off track. He finishes his aside with the atypical words *"Whereof enough ..."*, and concludes where he left off. He finishes by explaining that the putrefied body is brought back to life by the vapors observed in the upper part of the vessel, resulting in a *Cœlestial Substance* – the observation of which indicates that the synthesis is progressing properly:

> *And this is principally to be noted for a conclusion of this discourse, that there is a Cœlestial Creature generated, whose life is preserved by the Stars, and fed by the four Elements, which ought to be killed, and then putrified, which done, the Stars by means of the Elements will again infuse life into those putride bodies, that it may again be made that heavenly substance, which had its habitation in the highest Region of the Firmament, if that be done, you shall perceive that the Terrestrial is taken from the Cœlestial, with body and life, and that Terrestrial Body is reduced into a Cœlestial Substance.*

KEY VIII – PUTREFACTION OF ADAM

Valentine routinely paralleled Judeo-Christian mysticism with alchemical counterparts:

> *Know that when Adam the first man was formed by the great Creator out of a lump of Earth, there did not as yet appear any perceptible motion of life, until God breathed a Spirit into him, then was that lump of Earth endowed with power. In the Earth was Salt i.e. the body, the inspired Air was the Mercury, the Spirit. The Air by this inspiration did give a genuine and temperate heat, which was Sulphur, i.e. Fire, then it moved it self and Adam manifested by this motion, that a living soul was inspired into him …*
>
> *So Adam was first brought forth, generated and compounded of Earth, Water, Air, and Fire, of Soul, Spirit, and Body, and of Mercury, Sulphur, and Salt.*
>
> *For when Adam was Created, he was dead, and had no life of any virtue; but as soon as the operating, quickening spirit entred into him … In everything therefore, both Natural and Supernatural are copulated as one, and joyned together in their habitation, that every thing may be perfect.*

Spirit in Greek (*pneuma*), Latin (*spiritus*) and Hebrew (*ruaḥ*) each mean breath. *Soul* in Sanskrit (*atman*), Greek (*psyche*), Latin (*anima*) and Hebrew (*neshama*, meaning *Soul*, is closely related to the word *neshima*) also means breath. *Breath*, *soul* or *spirit* in the above sampling of passages, liken creating the Stone to the creation of Adam or man:

> *Now the performance of this composition is likened to the generation of man, whom the great Creator most high made …* – Morienus
>
> Genesis 2:7 *… then the Lord God formed the man of dust from the ground and breathed into his nostrils the breath of life, and the man became a living creature.*

BOOK 2 – CONFECTING THE STONE OF THE ANCIENTS

> *1 Corinthians 15:44 It is sown a natural body; it is raised a spiritual body. If there is a natural body, there is also a spiritual body. ⁴⁵ Thus it is written, "The first man Adam became a living being"; the last Adam became a life-giving spirit.*

Basil Valentine's *8th Key* encrypts the visual indicators for the blackening stage known in alchemical terms as *putrefaction*. According to Valentine, the product should decompose to a black mass and appear dead or inert for some time, accompanied by vapors and condensation appearing in the upper part of the vessel. He likens the process to the alchemist (Creator; God) infusing new life (Breath) into the first man (Adamic Earth) resulting in a type of spiritual re-birth (Cœlestial Creature) and sign of certain future success. A series of color-indicators follow thereafter signaling either transition or stabilization and the necessity to increase temperature accordingly, the subject addressed next in Valentine's *9th Key*.

KEY VIII – PUTREFACTION OF ADAM

Symbolism of VIII Clavis

- **Archangel Gabriel** = messenger of divine news, praise and glorification
- **Sheaf of wheat** = In art symbolizes the Eucharist, Last Supper, Passion and resurrection
- **Sowing of wheat** = penitence, modesty, as are hands over private parts
- **Wheat under the head** = harvest, fertility, resurrection
- **Grave** = bodily death and putrefaction, and future resurrection (grave-digger's arms raised upward)
- **Crows or ravens** = associated with Apollo and prophecy, St. Oswald, luck and protection (of the seed)
- **Twin archers** = Apollo as god of the sun, prophecy, archery light and truth with his twin Artemis/Diana
- **6 arrows and bull's-eye** = Keys 1 thru 6 coming progressively closer, with Key 7 hits the bull's-eye prophesizing success at confecting the Stone
- **Key above the bull's-eye** = The Key to the Kingdom of Heaven (finished Philosophers' Stone)
- **Walled enclosure with 12 gateless gates** = sealed environment, Kingdom of Heaven lies outside the enclosure, the gates represent doors opened by The XII Keys
- **Golubets** - 'pented' (or roofed) crosses = new home for the soul of the dead indicating the current (unroofed) dwelling as black/putrefaction and 4 additional ones 1) white stone, 2) red stone, 3) multiplied, and 4) projected

Key IX – Planetary Colors

... the Earth hath likewise seven Planets in it, which are brought forth and wrought upon by the Seven Heavenly Planets, only by a spiritual impression and infusion; and in this manner all the Minerals are wrought by the Stars.

Basil Valentine

Basil Valentine's 9th *Key* encrypts knowledge of the color indicators that mark stages of transition, stabilization and the need for temperature adjustment. The primary colors are black, white and red in specific order and indicate stabilization. Secondary colors according to Valentine's descriptions are grey, yellow and multi- or all colors:

> Saturn *the highest of the Cœlestial Planets, hath the meanest authority in our Magistery, yet is he the chiefest Key in the whole Art ... he must be reduced to the lowest light of all, and by corruption must be brought to a melioration, whereby* the black must be changed into white, and the white into red; *and the* other Planets must pass through all the colours *in the world, until they come to proper super-abounding tincture of the triumphant King.*

When Valentine uses the phrase, *"... in the whole world"*, he is using alchemical double-speak to mean ... *in the entire composition*, or alternately ... *in the reaction vessel*. He is attempting to explain to the reader that the *son of Saturn*, antimony, is not held in high regard by most. However, the fluidity or liquid nature of the composition of its pure *Essence* ($SbCl_3$), when added to a *fiery Metalline body* ($Au \cdot Sb_2O_2SO_4$), may be converted to a fixed impalpable powder as the finished Philosophers' Stone. He refers to *butter of antimony* as being *insensibly cold*, possibly in reference to the fact that it freezes at room temperature, that it resembles ice crystals upon absorbing moisture from the air or to contrast it with the *fiery Metalline* body. Alternately, *butter of antimony* was often referred to as *glacial* or *icy* oil of antimony. The resulting potential for *Transmutation* from base metal into alchemical gold is based on this unique composition derived from the *Tria-Prima*, flux-gold-antimony respectively.

> *Saturn be esteemed the meanest in the whole world*, yet hath he in him that power and efficacy, that *if his pure Essence*, which is beyond measure insensibly cold, *be added to a current fiery Metalline body*, its running quality may be taken away, and may be a malleable body, as Saturn itself is, but of far greater fixity, which Transmutation hath its original beginning and end from Mercury, Sulphur and Salt.

Saturn is presented as the *Prime Material*, antimony, in very figurative language in Valentine's *On the Great Stone of the Philosophers*:

> *I Saturn, the highest Planet in the firmament*, protest before you all my Lords, that I am the most *unprofitable and contemptible body, of a black colour*, obnoxious to the injuries of many afflictions in this miserable world, yet *am the examiner of you all*. For I have no abiding place, and *I take with me whatsoever is like unto me* …

The color sequence is carefully concealed by Valentine, yet matches the basic color-regimen found in most recipes that address the archetypal recipe for confecting the Stone. The primary colors are always the same and occur in a precise order – black → grey → white → yellow → red. These indicate stages of stabilization and temperature regulation requirements.

> *From Saturn proceed many colours*, that are made by preparation and Art, as *black, ash-colour, white, yellow, and red* …

Prior to examining Valentine's color regimen in detail, it is important to grasp his hidden meaning concealed in the form of a parable about transference of governing power:

KEY 9 – PLANETARY COLORS

Understand that one Planet must drive out and dispossess another *of his government, office, possession, and power,* until the best of all attain the highest power, *and* with the best and most fixed colour *given them by their Mother ... For the old world passeth away, and the new is come in its place, and* one Planet destroyeth another spiritually, *that* he that is strongest continues till the last by feeding upon the other, *two or three being overcome by only one.*

Planet	Administrative Role	Flag Carrier	Flag Color	Flag Image	Image Attire
Saturn	Master President	Astronomy	Black	Faith	Yellow and Red
Jupiter	Grand Marshall	Rhetoric	Ashen	Hope	Splendid colours
Mars	Warlike affairs	Geometry	Bloody	Fortitude	Red garment
Mercury	Chancellor	Arithmetic	All	Temperance	Glorious colours
Moon	New government	Dialectics	White Shining	Prudence	Silver Sky Blue
King Sun	Vice-Regent of the Kingdom	Grammar	Yellow	Justice	Golden
Queen Venus	Power in the King's Court	Music	Crimson	Charity	Green

According to the color-regimen scheme as Valentine presents it, the bloody red of Mars initially appears out of place. Yet Valentine's odd placement of Mars makes sense if we understand the accompanying description *"... and beareth rule in fiery heat ..."* as an indication to increase temperature. Indeed, it is this precise point in the experiment where a strong temperature increase is required:

Mars is hardened in warlike affairs, and *beareth rule in the fiery heat* ...

BOOK 2 – CONFECTING THE STONE OF THE ANCIENTS

His point regarding Mars' function of cutting and burning is made clear in *Of the Great Stone of the Philosophers*:

> *Then came Mars with his naked Sword* variously coloured, like a fiery glass, shining with divers and strange rays, *he brought this Sword to Vulcan the Gaolor*, to put therewith in execution all those things commanded by the Lords, which *when he had killed Mercury, he burnt his bones in the fire*, wherein Vulcan the Gaolor was very obedient.

The second of three anomalies unique to Valentine's scheme is the color attributed to Mercury. Where it should read *White*, he lists *All*, but this is simply Valentine cleverly indicating that Mercury is *Philosophers' Mercury*, its chemical identity being antimony, which can take on all colors. He makes this point explicitly clear in *The Triumphal Chariot of Antimony*:

> *Antimony, like Mercury is* comparable to a circle, without beginning or end, *composed of all colours*; and the more is always found in it, the more diligent and prudent the search which is made.

> He (the Artist; alchemist) can colour it red or yellow, white or black, according to the way in which he regulates the fire, since *Antimony, like Mercury, contains within itself all colours*.

> *In Antimony you will find all colours*, black, white, red, green, blue, and an incredible number of mixed tints (besides grey and yellow), which must all be severally known, and used in their own proper order.

The third anomaly unique to Valentine's scheme occurs in a odd passage where he explains that *Luna* (Moon) takes over office from the Queen with the help of (Philosophers') *Mercury*, both ultimately over-ruling the Queen. Some alchemists view gold as the King of alchemy, while others value the Philosophers' Stone above all alchemical substances. For Valentine it was neither gold, nor the Stone, but rather antimony that was the one *Triumphant Vehicle* of alchemy and all its permutations. Antimony is associated with Saturn, but also Luna and Mercury and

KEY 9 - PLANETARY COLORS

therefore holds the highest position in his alchemical hierarchy as usurping power from the King and Queen:

> And because Luna's [Butter of Antimony] Husband dyed she gained the Office her self, lest Queen Venus [the Stone] should get into the government again; for she called her to an account of her office, then the Chancellour [gold-antimony glass] assisteth her, that *a new Government may be established, and both them rule above the Queen…*

However, this is not what Valentine was attempting to communicate. He was hinting at the process of *Multiplication* detailed in the *11th Key*, whereby the Queen (Philosophers' Stone) is combined with additional gold-antimony glass and equal measure of *butter of antimony* and the entire synthesis repeated. The *multiplied Stone* was viewed by alchemists as being superior or of a higher order than the un-multiplied.

He understood the color-regimen for confecting the Stone as incorporating each of the seven planetary influences, yet the primary color progression always seems to remain:

black → (transition) → white → yellow → red

In some of the important *Keys*, Valentine includes a diversion in an effort to throw the unworthy aspirant off track – a type of alchemical red herring. Each diversion is a valid alchemical point or topic that appears to be out of context. He again displays an acute understanding of Greek mythology and astrology in the *9th Key*'s distraction, which occurs in the form of instructions to rearrange the signs of the zodiac into two categories in a balance, as if on a scale:

> *For a final conclusion you may understand hereby that you must take the Cœlestial, Libra, Aries, Taurus, Cancer, Scorpio, and Capricorn, and at the other end of the balance put Gemini, Sagittary, Aquaries, Pisces, and Virgo, then cause that the golden Lion, leap into the lap of Virgo, so will that part of the Scale be the weightiest, and weigh down the other…*

BOOK 2 – CONFECTING THE STONE OF THE ANCIENTS

He begins with Libra as a hidden clue that two columns, described as scales, are required. He leaves Leo out of the discussion until the finale, instructing the operator to cause that so far unnamed *golden Lion to leap into the lap of Virgo* – suggesting that Leo is on one side of the balance but then leaps onto the other.

Valentine's imagery becomes clear upon doing as he instructs:

Animals				Humans / Gods		
♎ Libra	(Scorpion's Claws)	♎		♊ Gemini	Twins	♎
♈ Aries	Ram	△		♐ Sagittarius	Archer	△
♉ Taurus	Bull	▽		♒ Aquarius	Water-bearer	♎
♋ Cancer	Crab	▽		♓ Pisces	(Aphrodite & Eros)	▽
♏ Scorpio	Scorpion	▽		♍ Virgo	Virgin Maiden	▽
♑ Capricorn	Goat	▽		♌ Leo	Lion	△
~~♌ Leo~~	~~Lion~~	△				

Earth-Water Dominant		Air-Fire Dominant
Mostly Descending		Mostly Ascendant
2▽ 2▽ 1♎ 1△		1▽ 1▽ 2♎ 2△

According to Valentine, the ability to enact this feat alchemically upsets the cosmic balance. The alchemical imagery of a *Lion* and a *Virgin* features in his woodblock prints in *Keys 5* and *11*, obviously symbolizing union between the *Lion-King* as gold and the *Virgin-Queen* as antimony or *butter of antimony*, but also uniting animal royalty with human in this scenario. The remainder of the passage demands explanation.

By having Leo join the human side, a balanced imbalance occurs. Libra appears out of place until remembering that Libra was considered the *Scorpion's Claws* in Greek astrology, and thus is appropriately placed with the animals. Another obvious objection to this scenario is that the sign Pisces is symbolized by fish, yet Valentine again reveals his fluency in mythology. He correctly places Pisces in the Human-God category because according to Greek myth, Aphrodite (Venus) and her son Eros (Cupid) transformed themselves into the two fish of Pisces in order to escape the wrath of the "Father of All Monsters", Typhon. From an alchemical view, although the four elements are in perfect balance, the Human side of the scale, with the *Lion in the lap of Virgo*, now possesses the Stone and is therefore the *"weightiest"* due to its influence.

The entire riddle in the form of a zodiacal-elemental yin-yang balancing exercise was apparently designed to symbolize a conjunction or union of the four elements – a creative conceptual metaphor for confecting the Philosophers' Stone. The animal side of the scale symbolizes a 2:1 ratio of earth-water : air-fire, whereas the other side is balanced with 2:1 ratio of air-fire : earth-water. Exercises such as these, products of the creative mind behind the pseudonymous writings of Basil Valentine, indicate masterful conceptual blending. In the passage below, Valentine juxtaposes the 12 signs of the Zodiac with the 7 planetary bodies, yet cleverly disguises his doing so with the following passage:

> ... then let the twelve Signs of the Heaven come in opposition to the Pleiades.

The Pleiades in mythology was known as the 7 sisters and were visible in the night sky, in other words seven celestial related bodies as a cover-term for the seven planetary bodies visible from earth with the naked eye. Alternately, he may have been alluding to Hermes of Greek religion and mythology, son of Zeus (of the 12 Olympians) and the Pleiad Maia. In any case, Valentine was balancing stellar and planetary correspondences to symbolize confecting the Philosophers' Stone. In his closing statement, he mentions *finishing all the colors of the [microcosmic] world* in reference to the planetary colors detailed earlier in the chapter. The last and greatest color-duo is attributed to antimony as gloriously shining all-in-one Luna-Mercury, governor of the kingdom overtaken from the dispossessed King and Queen. For Valentine, antimony was the *All-in-One*.

> *And so after the finishing of all the colours of the world, there will at last be a conjunction and union, that the greatest cometh to be the least, and the least to be the greatest,*
>
> *If that the nature of the whole world remained,*
> *Only in one state, form, or quality,*
> *And other forms could not by Art be gained,*
> *The wonders of the world would cease to be.*
> *And Natures mysteries would not be raised,*
> *For whose discoveries let God be praised.*

Basil Valentine's *9th Key* encrypts his perspective on the matrix of interconnected colors, planetary bodies and zodiacal star signs as they pertain to confecting the Philosophers' Stone. His complex abstractions are best approached from a frame of reference that includes practical alchemy, Greco-Roman mythology, astronomy and astrology and court intrigue. The real and enduring value of the *9th Key* to practical alchemists is the focus on the primary color-indicators depicted in avian form fitting his color scheme and immortalized in his woodblock print. Each of the primary colors – black, transition, white and red – symbolized as a crow, peacock, swan and phoenix respectively, are points in the synthesis where the reaction has stabilized or must be further catalyzed by an increase in temperature, the subject addressed next in Valentine's *10th Key*.

KEY 9 – PLANETARY COLORS

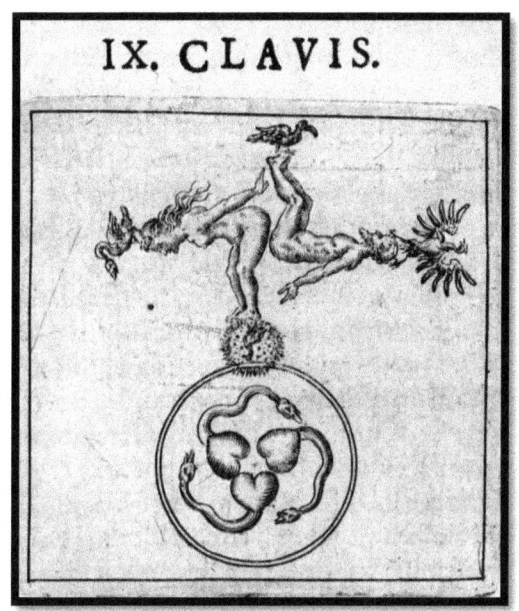

Symbolism of IX Clavis

- **Naked Adam** = king, male, solar, fiery earth principle, pure ($Au \cdot Sb_2O_2SO_4$)
- **Naked Eve** = queen, female, lunar, airy water principle, pure ($SbCl_3$)
- **Womb germination of Tria Prima** = serpent seed germination as salt, sulfur and mercury unite in mutual love
- **Black raven/crow** = sophic decomposition, blackening, melanosis / nigredo, prophecy
- **Peacock** = vegetative germination, transition and transformation, rebirth, knightly vigilance, Christian watchfulness
- **White duck** (swan) = white sulfur, volatizing by liquefying, marriage, purity of soul, everlasting life
- **Phoenix** = resurrection and eternal life
- **Antimony** = as "son of Saturn", outline of the entire picture, ☿, "... chiefest Key of the whole Art" and primary indicator that Valentine is addressing the archetypal Alexandrian-style "Great Stone of the Ancient Philosophers" and not a solvent-vehicle, acetate tincture or mercurial style of alchemy

Key X - Water, Ashes and Sand

*Having thus attained the Matter, nothing remains
but that you look well to the Fire, that you observe its Regiment,
for herein is the highest concernment, and the end of the work ...*

Basil Valentine

Basil Valentine's *10th Key* encrypts knowledge of the heat regimen necessary to bring not only the Stone to completion, but also fixation to a fireproof state. In the opening paragraph, Valentine reveals that the Stone he has been describing is the archetypal *Great Stone of the Ancient Philosophers*, the *Tincture of the Philosophers* as a legacy inherited from other alchemists long before him. He refers to the archetypal Philosophers' Stone by the cover-term *body of Seven*:

> In *our Stone made by me, and others long before me*, are all the Elements, and all the mineral and metalline forms, yea, and *all the qualities and properties of the whole world contained*; for therein is found the greatest and strongest heat: For *by its great internal fire the cold body of Seven is warmed*, and by that heating is changed into the best gold: In it also is found the greatest cold, for by its conjunction the hot Nature of Venus is temperated, and quick Mercury coagulated, and by the same reason, by its fixity transmuted into the best fixt Gold ...

In the passage above, Valentine touches on the practical application of the Stone for transmuting copper (Venus is temperated) or mercury (quick Mercury coagulated) to alchemical gold, yet to do so, informed knowledge concerning temperature regulation is necessary to bring the Stone to completion:

BOOK 2 – CONFECTING THE STONE OF THE ANCIENTS

> Because *all these properties of our matter of the great Stone* are infused by Nature, which properties *are concocted and maturated by the degrees of fire*, untill they have attained the highest perfection ...

Temperature Regulation

Having detailed the color indicators in the previous chapter, Valentine continues with a discussion of temperature regulation beginning with low, yet ascending heat.

1. The Matter is dissolved in Balneo [40-42 °C],
2. [heat raised to the temperature of horse dung; 47-50 °C]
3. and united by putrefaction in Ashes it produceth flowers [70-80 °C];
4. in Sand all its superfluous humidity is dryed away [100-110 °C],
5. but a quick fire maturateth it with fixation [120-130+ °C],

> ... not that you must needs use *Balneum Maria*, *Fimus Equinus* [horse dung], *Ashes* and *Sand* successively, but that the degrees and regiment of the Fire be so performed: For the Stone is made in an empty Furnace, of a threefold defence firmly lockt up, inclosed and concocted with a continual Fire, until all Clouds and Vapours vanish, and the Garment of Honour appear in the greatest splendor ...

The reaction occurring inside the vessel is controlled by temperature more so than by duration. It is possible to force the reaction by raising the temperature one degree per hour, yet alchemists did not operate in this manner. Instead, they began with the lowest temperature, often described as the heat of one's hand or of a hen sitting on her eggs – the temperature required to catalyze *decomposition* (blackening). The second adjustment is typically described as the temperature of horse-dung, which equates to roughly a 5-10 °C increase – the temperature required to catalyze *vegetative germination* (transition). The third adjustment is described as an ash bath, which is approximately a 20-30 °C increase – the temperature required to bring the product to a molten state (whitening). This is followed by an additional increase of around 20-30 °C – the temperature required to vaporize any liquid and to dry the product (reddening).

> And *when the King can lift up his Armes no longer, he hath obtained the government of the whole World, for his is made the King of everlasting fixity*, no danger can ever hurt him, for he is become invincible ...
>
> Now let me tell you, when your Earth is *dissolved in its proper Water, dry away the Water th[o]roughly by its due Fire, then will the Air breath into it a new life*, and whom this life is incorporated, *you have a Matter*, which deservedly can have no other name, than *the great Stone of the World* ...

The passage above signifies a fixation or maturation stage brought about by allowing the product to cool, opening the vessel and heating once more to what is described as a quick fire, which equates to 130 °C or higher – the temperature required to thermally decompose the product to its fixed or mature form. Valentine introduced the importance of temperature regulation in the 7th *Key* and in the 10th *Key* presented it in detail, with an emphasis on the annealing or fixation step. Increasing the temperature gradually over a period of approximately 3 days results in the Philosophers' Stone as Valentine describes it. It will decompose and liquefy before finally solidifying:

Another phenomenon observed by Valentine is the color-change that occurs as the product cools. Towards the end of the reaction, the product will become quite red in color, but this color changes to a darker color that can appear reddish-purple or even tending towards brown and is dense due to its gold content.

> *Its colour declineth from a shining redness to a purple*, from a Ruby to a Granate [or maroon; dark crimson] colour, and in weight it is exceeding and very great.

BOOK 2 – CONFECTING THE STONE OF THE ANCIENTS

Maturing and Fixing in the Fire

Earlier in the chapter, and somewhat out of place, he hints that a very real potential danger exists in the form of hastily attempting to work with the Stone before it is fully matured:

> *If fruits are gathered before they are ripe, they are untimely and unprofitable, neither are they for use; so unless the Potter burn and concoct his wares enough in the fire, they are unfit for use, because they were not sufficiently Maturated in the fire.*
>
> *So also concerning our Elixir you must diligently consider that a just time be given it, and that before that time nothing of its virtue be detracted, lest it be aspersed and esteemed for an unworthy thing.*
>
> *For in truth, if our Stone be not sufficiently maturated, no ripeness can be produced from it.*

Valentine is alluding to the process of *maturation* or *fixation*, which essentially requires heating the finished Philosophers' Stone product quite hot to temper or anneal it, resulting in its being considered fireproof. Valentine makes this point clear early in the chapter, suggesting a level of importance. Annealing is a heat treatment that alters the structural and sometimes chemical properties of matter in both metallurgical and glassmaking processes. Subjecting the product to heat removes excess moisture along with trapped gasses. Valentine touches on its importance as part of confecting the Philosophers' Stone in *On the Great Stone of the Philosophers*:

> *Whosoever desireth to know what the All in All is, let him make very great Wings for the Earth, and force her so much, that she lift her self up, and raise her self on high, flying through the Air into the Supream Region of the highest Heaven. Then burn her wings with a very strong Fire, that the Earth may fall headlong into the Red Sea, and be drowned therein, and with Fire and Air dry up the Water, that thereof Earth may be made again; Then I say have you All in All.*

His instructions, *with Fire and Air dry up the Water*, is a subtle clue that this final annealing process is performed once the vessel is opened or even possibly once the product is removed.

Supporting Documents
In the supporting text *The Manual Operations of Basil Valentine, The Preparation of the Great Philosophick Stone*, he explains that the purpose of finishing the Stone is to prepare it to be *the greatest Medicine* by removing all corrosives:

> But observe, that, as you will hear afterwards, all Corrosives must be washed away again from it [the Stone] and separated, so that our Stone may be severed from all poison, and be prepared to be the greatest Medicine. ... For considerable is both the beginning of the work, and the work it self: but the end is high and excellent, all which knowledge and experience will discover and bring to light.

Valentine's conclusion makes it perfectly clear that the 10th Key signifies completion of the process for confecting the Philosophers' Stone according to his methodology. The obligatory praise and reverent display of gratitude closes Valentine's instructions for creating the archetypal Philosophers' Stone as revived in the European alchemical tradition.

> Whosoever shall be adopted to the Stone, let him return thanks to the Creatour of every creature, for that Cœlestial Balsame; and let him pray that for himself and his neighbor he may use it for the sustentation of this temporal life, and that he may enjoy eternal happiness in this valley of miseries, and in the other world to come,

> Let God be highly praised for this his unexpressible gift and grace forever, Amen.

Basil Valentine's 10th Key revisits the temperature regimen in somewhat more detail than is provided in the 7th Key. It suggests a correlation between temperature and the primary and secondary color indicators explained in the *9th Key*, yet Valentine leaves it to the operator to determine precisely how the two regimens correlate. As indicated in the woodblock print, he views the entire process as Antimony passing through the heart of the seven heavens in a precise hierarchical order.

Initially the product appeared as a dead matrix of three specific ingredients (Luna, Mercury, Sol), yet the *Wise* knew how to bring about anastasis or resurrection of the product, reborn as a homogenous new and exalted substance baptized by fire and impervious to it. At this stage of working with the Philosophers' Stone, two options remain – 1) the stone may be increased in volume, or 2) applied industrially towards the production of alchemical gold, the subjects addressed next in Valentine's *11th* and *12th Keys*.

KEY X - WATER, ASHES AND SAND

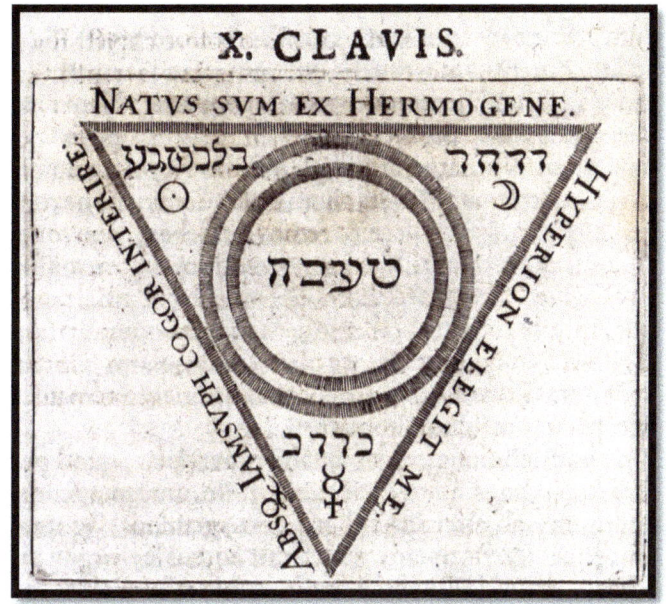

Symbolism of X Clavis

- **Right** ↙ = Hyperion [father of Helios (Sun), Selene (Moon) and Eos (Dawn)] has chosen me.
- **Left** ↖ = Without already possessing wisdom, I am destined to perish:
- **Top** → = I am [re]born of Hermes, the homogenous.
- **Center** = טֶבַע (Nature, Substance, Element) [of 3 ingredients]
- **Luna** = יְרֵחִי (Lunar)
- **Mercury** = כררכ (carefully through; hierarchical; the hierarchy of …)
- **Sol** = בלֵב שבע (heart of seven)
- [Antimony as] Luna has carefully transitioned through the heart of seven [governors, i.e. colors, planets, phases, etc.]

BOOK 3

Multiplication
&
Fermentation

*Now as this excocted and perfect substance
is the highest, chiefest, and greatest* Universal Medicine *unto man,
even so on the other side the same matter after its* fermentation,
is a Tincture also, and the chiefest, greatest, and most powerful
Universal Medicine *upon all Metals ...*

Basil Valentine

Key XI - Noble Offspring

*... then is that mighty generation nourished with his own flesh,
and is renewed with his own honorable blood;
if you have done this rightly you shall leave a numerous generation ...*

Basil Valentine

Basil Valentine's 11th *Key* encrypts the alchemical process for augmenting or multiplying the Philosophers' Stone, which means to increase the amount of the Stone in mass and volume. Valentine does not speak directly however, and the process is presented as a riddle in the form of a parable:

> *The eleventh Key of the multiplication of our Great Stone I will discover and reveal unto you by way of Parable after this manner.*
>
> *In the East there dwelt a Knight, called Orpheus, who mightily abounded in wealth, and did excell in all good things: He chose and took for his Wife his own Sister Euridice; but when he could have no issue by her, he imputed it to his sins in chusing his own Sister for his wife: With his daily prayers he besought and beg'd the most high God, that he would communicate to him his Grace, and give way to his request.*

To make any progress decoding the riddle, it is necessary to attribute chemical identities to each character in the story. So far, he has introduced a wealthy knight named Orpheus and his sister-wife Eurydice. The story of Orpheus and Eurydice (spelled *Euridice* by Valentine) is well known in Greek religion and mythology, yet Orpheus is typically depicted as a musician, poet and prophet, which is to say that he is a lover rather than the battle hardened warrior knight portrayed by Valentine. More importantly, Orpheus was known to all Greeks as being associated with the Orphic mysteries, which held that the human soul is both divine and eternal, but subject to transmigration as reincarnation until ultimate liberation from the cycle upon achieving perfection. In the parable, Orpheus represents antimony, the eternal vehicle of alchemy that can assume so many different forms and colors. His sword is antimony's ability to cut or destroy other metals. The wealth of Orpheus is the gold in his possession. The chemical identity for the character Orpheus and his wealth is gold-antimony oxysulfate (gold-antimony glass; $Au \cdot Sb_2O_2SO_4$).

Eurydice as a cover-name for alchemy's secret solvent is derived from the teachings of Morienus al-Rumi, whose writings were widely known to European alchemists during the period in which the Valentine papers were written. Valentine recycled no small amount of alchemical cover-terms, imagery and symbolism directly from Morienus' alchemical classic *The Composition of Al-Kīmyā'*, composed during the 7th century and one of the earliest alchemical texts translated to Latin. Morienus used the cover-name Eudica (Eurydice) to indicate *butter of antimony*.

A new character is introduced that again reflects Valentine's deep grasp of Greek mythology, a flying, hot, wise old sage who explains to Orpheus the process to successfully breed new offspring with his wife – the secret being a mixture of three blood extractions.

> *Being sometime overcome with deep sleep, there came to him a man flying named Phœbus, he [Orpheus] toucht his feet, which were very hot, and [Phœbus] said [to Opheus], Most Noble Hero, you have travelled through many Towns and Regions, and have undergone many dangers in the vast Ocean, and have sustained so much of the war, that you have acquired that Noble Order, and have merited that dignity before any other, having broken many weapons in Duels and Tournaments, and have often obtained honour by the Venerable Matrons: Therefore my Father in Heaven commanded me, that I should declare unto you, that your supplications were heard; therefore …*

KEY XI – NOBLE OFFSPRING

Students of Greek mythology will immediately recognize Phœbus to be the Roman version of Apollo, the light-bringer, sun god whose mother Leto was associated with a wolf. In Latin, his name means "radiance", yet in the Macedonian tongue of Alexander the Great, Apollo is pronounced *pella* (πέλλα), meaning "stone". Here Valentine was referring to the Philosophers' Stone by the cover-name Apollo, who just happens to be father to both Orpheus and Eurydice in Greek mythology. The parable portrays a father giving advice to his son, with heavy overtones of blood relations:

1. ... *you are to take the* blood of your right side *[gold-antimony glass],*
2. *and the* blood out of your Wife's left side *[butter of antimony],*
3. *and the* blood which was concealed in the heart of your Father and Mother *[the Philosophers' Stone], they are naturally* two, and yet one blood,
4. conjoin these [3 bloods] together, *and cause gain that* they enter the Globe of the seven Wise Masters [planetary governors],

The *seven Wise Masters* is Valentine's cover-speak for the seven planetary governors who rule by their influence, (color and heat indicators) and thus, the Multiplied Stone is brought to maturity through the entire color and heat regimen. The *Globe of the seven* symbolizes the reaction vessel. Thus far, he has cleverly revealed the substances and equipment necessary to increase the Stone.

> ... nakedly enclosed; then is that mighty generation [Apollo- Phœbus] nourished with his own flesh [of son and daughter], and is renewed with his own honorable blood; *if you have done this rightly you shall leave a numerous generation ...*

The process of Multiplication was first clearly described by Morienus during the 7th century and recorded in *The Composition of Al-Kīmyā'*, yet he termed the process *fermentation*.

95

BOOK 3 – MULTIPLICATION & FERMENTATION

> *Then take such a quantity of the elixir [Philosophers' Stone] as to form eleven parts for every ten parts of the white body [flowers of antimony]. Mix these, and for every ounce of this mixture, add one-fourth of a dram of Eudica [butter of antimony]. – Morienus al-Rumi*

Although Morienus' methodology is slightly different from Valentine's, it demonstrates the origin of the process, yet also speaks for the experimental and innovative nature of Valentine and other European alchemists who endeavored to optimize alchemical processes.

Morienus provides instruction as regards measures, while Valentine does not. The chemistry however reveals that regardless of the amount of the original Philosophers' Stone to be multiplied, an equal measure of the two compounds (*gold-antimony glass* + *butter of antimony*), or with butter of antimony in very slight excess, is optimal.

> *If you do this often and always beginest anew, you shall see your Childrens Children: That the great World shall be th[o]roughly replenished by the generation of the lesser, that may be abundantly possess the Cœlestial Kingdom of the Creator.*

He then proceeds to explain in the form of a parable that everything was a success. In the manner typifying an authentic alchemical text, Valentine plainly states that he left absolutely nothing out, and if he is not fully understood, it is the fault of the aspirant-interpreter and none of his own.

> *Now, Son of Art, if you have understanding, you need no other interpretation; but if you have no understanding impute it not unto me, but to your own ignorance.*

Operative alchemists familiar with Alexandrian archetypal alchemy through Latin translations of Greek or Arabic sources would have been able to decode and confirm that Valentine was not only in possession of the Stone, but also fluent in alchemical language and well versed in Greco-Roman mythology and Christian mysticism. He is addressing not

KEY XI – NOBLE OFFSPRING

only adept-alchemists in the passage below, but also aspiring adepts on the path to discovery:

> *I have described the whole process figuratively, and after the Philosophick manner, and as my predecessors have done, yea, and more plainly than them, for I have concealed nothing:* If you remove the veil from your Eyes, you shall find that which many have sought, and few find, for the Matter is absolutely expressed by its Name, the beginning, middle and end also demonstrated.

BOOK 3 - MULTIPLICATION & FERMENTATION

KEY XI – NOBLE OFFSPRING

Symbolism of XI Clavis

- **Orpheus** = god of the mysteries, son of Apollo, magician, could charm even stones with his music
- **Eurydice** = or Argiope, daughter of Apollo and wife of Orpheus, died by serpent's venom (Ios, tincture)
- **Mature lions** = The Finished Red Lion and the immature Green Lion as essentially the same composition
- **Lion cubs** = offspring, successive batches of the Stone made via the multiplication process
- **Lion devouring another** = the Finished Stone (Red Lion), reacting with the unfinished (Green Lion) in a courting / mating ritual
- **After-birth** = indicates "blood of the lion", or alternately "offspring or bloodline"
- **Flowers of gold and antimony** = the Finished Stone and Multiplied Stone are of the same composition
- **Sword of Orpheus** = Chalybs (Caliburn), butter of antimony, sword of wisdom

Key XII – Fermented Tincture & Fixt Gold

*... then are the pure Gold and the Stone made a meer Medicine,
of a subtile, spiritual, and penetrating quality:
For without the ferment of Gold the Stone cannot operate,
or exercise its tinging quality ...*

Basil Valentine

Basil Valentine's *12th Key* encrypts the alchemical process for applying the Philosophers' Stone to the creation of alchemical gold. When the Philosophers' Stone contains a precious metal for the purposes of creating alchemical gold or silver, the Stone is known as the *Tincture*, or more accurately, *Tincture of the Philosophers*. This term is the most ancient name for the Philosophers' Stone and is derived from the Greek *Ios*, meaning *tincture*, *poison* or *venom*, the *poisoned arrows of Heracles* or *purple* to designate royalty. Heinrich Khunrath referred to this variant of the Stone as the *Lesser* or *Specific Stone*, whereas Paracelsus referred to it as the *Tincture of the Philosophers*. The process of using the *Tincture of the Philosophers* and the nature of alchemical gold is detailed in *Cracking the Philosophers' Stone*.

Valentine begins the *12th Key* with an allegorical tale about a fencer who does not know how to use his weapon, as paralleling an alchemist who does not know how to use his *Tincture* – "*He that hath well attained the Mastery ... wins the Prize*":

> *So he that hath by the Grace of the Omnipotent God obtained the Tincture, and knoweth not how to use it; so it happeneth unto him as was said of the Fencer, that knew not the use of his weapon:* **But seeing this twelfth and last Key is for the finishing of my Book, I will not detain thee any longer in parabolical or figurative expressions, but** *without any*

BOOK 3 – MULTIPLICATION & FERMENTATION

> *obscurity I will discover [reveal] this Key of the Tincture in a most perfect and true process ...*

What follows is Valentine's precise method for creating what is known in alchemy as the *Fermented Stone*. True to his word and without figurative expressions or any obscurity, he outlines the procedure in the *12th Key* in surprising detail:

> *When the Medicine and the Stone of the Philosophers is made and perfectly prepared out of the true Virgin's Milk:*
>
> 1. ... *[of the Stone of the Philosophers] take thereof one part,*
> 2. *of the best and purest Gold, melted and purged by antimony, three parts,*
> 3. *and reduce It [the gold] into as thin plates as possibly you can [gold-leaf],*
> 4. *put these together into a Crucible, wherein you use to melt Metals,*
> 5. *first give a gentle Fire for twelve hours,*
> 6. *then let it stand three days and nights continually in a melting Fire,*
>
> *... then are the pure Gold and the Stone made a meer Medicine [of metals], of subtile, spiritual, and penetrating quality: For without the ferment of Gold the Stone cannot operate, or exercise its tingeing quality, being too subtile and penetrative: but being fermented and united with its like ferment, the prepared tincture obtaineth an ingress in operating upon other bodies.*

For Valentine, the process of combining a portion of the Philosophers' Stone with gold would have been understood as *seeding* or *fermenting* the product with a bit of starter as the *seed* or *ferment of Gold*. This practice is common and fully in keeping with a number of artisanal craft-technologies such as brewing, baking, distilling and glassmaking traditions in which *pre-ferment* or *fermentation starters* such as baker's yeast or live yeasts play an important role. Valentine does not provide the reader with any clear description of the Fermented Stone's

KEY XII - FERMENTED TINCTURE & FIXT GOLD

appearance, nor does he describe any characteristic other than to suggest a method of application:

> Then *take of the prepared ferment one part, to a thousand parts of melted Metal* [copper or silver], *if you would tinge it, then know for a very certain truth, that it shall be transmuted into good and fixt Gold*:

The passage that follows is a very poetic description of an alloy. It is important to notice here that Valentine describes the process as *one body embraceth the other*, and then qualifies this by suggesting that the alchemical gold displays the nature of gold due to the original gold.

Nothing in the qualifying passage that follows can be construed as Valentine either affirming or denying transmutation of what might be described today as an elemental or nuclear sort, only that the resulting product is *like gold*, deriving the likeness from the original gold.

> For the *one body embraceth the other* although they be not alike, yet by the force and power added to it, *one is made like unto it, like having its original from like*.

There are strong arguments for and against European alchemists believing alchemical gold to be the equivalent of elemental gold, or alternately a species or unique type of gold alloy, made specifically by alchemical means, yet not identical to elemental gold. This topic is explored in detail and at length in *Cracking the Philosophers' Stone*.

And so, the adept-alchemist at this point, having fully deciphered Valentine's multi-layered mystery, is in possession of the *Keys to the Kingdom of Alchemy*, behind whose doors lies hidden not only the means to confect the *Stone of the Ancient Philosophers*, but to multiply, perfect, and ferment the *Tincture* for tingeing metals and/or making alchemical gold.

BOOK 3 – MULTIPLICATION & FERMENTATION

He that useth this means, to him is revealed all truth. The Porches of the Pallace have their goings forth at the end, and this Policy is not to be compared to any Creature: For it possesseth All in All, as naturally and originally in this world can possibly be done under the Sun.

O Beginning of the first Beginning, consider the end.
O End of the last End, see to the Beginning.

And let the Middle be faithfully remembred by you, then will God the Father, Son, and Holy Ghost, give unto you, whatsoever you require for Spirit, Soul, and Body.

KEY XII – FERMENTED TINCTURE & FIXT GOLD

Symbolism of XII Clavis

- **Alchemist** = man as creator with the power to affect transmutation
- **Furnace** = suggests that the entire process is a dry, or metallurgical method
- **Lion devouring serpent** = the ferment of Gold overpowering the serpent or antimony-content
- **Sun and moon in window** = Sol and Luna, as Gold and Antimony
- **Flowers and Mercury** = flowers of antimony as prime material and sophic mercury (1 part flowers for glass and 1 part for butter of antimony))
- **Two Flowers** = the twin Mercuries of the 2nd Key, Universal Stone and Lesser Tincture of the Philosophers
- **Tongs** = suggest skill in crucible fusion technique
- **Metallurgy tools on shelf** = Triangular Hessian and Conical German graphite crucibles, scorifier and cupel
- **Tools on workbench** = hammer, copper-plate, measuring utensils, wire-snips, scale and two books
- **Alchemist pointing at a post-medieval Hessian triangular crucible** = alchemical crucibles favored by alchemists, symbolize the Great Work of the Tria Prima, the Philosophers' Stone

APPENDIX

Cover-terms used by Valentine

Stibnite / Antimony (Sb_2S_3; Sb)
- Grey Wolf
- Son of old Saturn
- Regulus

Refined and Particle-Size Reduced Gold ($AuCl_3$; Au)
- Sol
- Sun (image)
- Honorable Mantle
- Rose of the Masters
- Red Blood of the Dragon
- King with his Heavenly Splendor
- Purple Mantle
- Cœlestial Sulphur
- Aes Hermetis
- Ferment
- Sand
- Red Fox (image)

Antimony Ashes and Glass ($Sb_2O_3 \cdot Sb_2S_3$ and $Sb_2O_2SO_4$)
- Ashes
- Cœlestial (principle)
- Clarified Salt
- Queen
- Diana
- Moon (image)
- Glassy Sea
- Dragon-Serpent

APPENDIX

Gold-Antimony Glass (Dry Compound; $Au \cdot Sb_2O_2SO_4$)

- *Adam*
- *Naked Adam (image)*
- *Earth*
- *(Soul of the) Earth*
- *New Man*
- *Man*
- *Double Fiery Man*
- *Orpheus (parable)*
- *Blood of [Orpheus'] Right Side*
- *King*
- *Lion's Blood*
- *Leo Viridis*
- *Philosophick Lion*
- *Golden Lion (riddle)*
- *Leo (riddle)*
- *Beast-Griffin (image)*
- *Seed*
- *Salt of Glory*
- *Salt of Ashes*
- *Salt of the Philosophers*
- *Sal Philosophorum (image)*
- *Glass*

Butter of Antimony (Liquid Compound; $SbCl_3$)

- *King's Palace*
- *Naked Eve (image)*
- *Mercurial Spirit*
- *Mercurial Water*
- *Mercurial Oil*
- *Mercury (II)*
- *Hidden Swordsman (image)*
- *Argent Vive*
- *Water*
- *Aqua (image)*
- *Starry Water*
- *Aqua Cœlestis*
- *Spiritual Water*
- *Salty Sea*
- *Aqua Permanens*
- *Magnet*
- *True Bird and Eagle*
- *True Bird [of Hermes]*
- *Eagle*
- *Swan*
- *Eurydice (parable)*
- *Blood of Wife's left side (parable)*
- *Queen (image)*
- *Virgin (riddle)*
- *Virgo (riddle)*
- *Moon (image)*

POSTFACE

Attempting an accurate interpretation of Valentine's *XII Keys* was a delightful challenge, and a difficult one. Perhaps the greatest consideration was whether to include instructions that occur in supporting texts or to address only those included in *The XII Keys*. It was ultimately decided to include supporting texts as correlating to the original document with full understanding that this approach may have resulted in an over-sophisticated interpretation. The substances and processes that occur in supporting documents, from a purely procedural perspective, reveal a much more complex story than do *The XII Keys* themselves if read separate and apart from other Valentinian texts. With the goal of elucidating Valentine's particular brand of alchemy in mind however, it seemed relevant to include the supporting texts if they in any way helped to illuminate the corresponding *Key* or a related procedure.

Apart from the above, several other lingering considerations must also be addressed. These include the omission of any form of therapeutic or human in-vivo application of the *Stone of the Ancient Philosophers*, the exclusion of any reference to increased potency in the 11th *Key*, as well as the conflict that arises in light of *Phalaja* and other "solvent-based" styles of alchemy familiar to Valentine, Paracelsus and other alchemists of the post-medieval period. Finally, since Valentine included gold as a primary ingredient throughout the text, the notion that iron or copper could substitute for gold was left for the Postface so as to avoid confusion.

Valentine is typically identified as the type of alchemist whose primary motivation was what might be described as pharmaco-alchemy. Yet the only direct reference to a "medicine" in *The XII Keys* is in the context of a medicine for impure metals, i.e. for the creation of alchemical gold.

Valentine, like so many other alchemists, certainly would have considered that the Stone, as Universal Medicine to heal "sick or impure metals", heals all bodies – including human. While this seems an odd omission for Valentine however, it actually reinforces the notion that he was addressing the archetypal *Chrysopœia* (gold making) of Alexandrian alchemy in his text.

Many classical alchemy texts refer to *multiplication* or *augmentation* of the Stone (addressed in the 11th Key) as an increase in both quantity *and* potency or virtue. The context in which Valentine uses the word *augment* in the 11th Key is in keeping with the standard definition "to increase", yet it can and has been interpreted as "to add potency". This potency-adding aspect of the Stone appears to be missing from *The XII Keys*. The notion of transmuting "10 parts of base metal, then 100 parts, then 1000 parts, then 10,000 parts, etc." appears to be a theme fully developed in European alchemy, yet nowhere in the original text does Valentine even hint that the Stone is made somehow stronger or more potent according to a graduated process. Valentine's process suggests *increase* only in quantity, i.e. mass and volume, which is to say that his 11th Key only instructs how to make more of the same substance. The 12th Key is where one would expect to find such instructions, but it is missing from that Key as well. The only potency addressed in any way is his claim that the *fermented stone* tinges 1,000 times its weight of base metal.

In the supporting documents, specifically the 3rd Key in *The Elucidation of the XII Keys*, the reader is introduced to what might be described as *solvent vehicle alchemy*. A *solvent vehicle* is a substance that dissolves a chemical, in this case gold trichloride, into a uniform solution without causing a chemical change or reaction. These solvents were known in European alchemy as *menstrua*, or by the singular *menstruum*, and were chiefly employed as *solvents* or *vehicles* for the preparation of alchemical medicines. The use of *solvent vehicles* was prevalent among Prussian,

German and Swiss alchemists such as Paracelsus, Valentine and others in their lineage.

Solvent vehicle alchemy is firmly based on wet chemistry precedents such as distillation, ethanol applications, mineral acid production and their use in dissolving metals, the technique of isolating and migrating metal salts using a *solvent vehicle* and combining the metal salt with another salt or ash such as potash, potassium hydroxide, lead acetate, etc. These prerequisites were developed during the period of medieval Islamic alchemy, which exhibit a unique and easily identifiable stylistic profile that became immensely popular with some European operators. The two most popular class of solvents used in this manner were either a form of ether (hydrochloric ether, chloroethane or diethyl ether) or acetone derived primarily from dry distillation of lead acetate.

Dr. Ahmad Yousef al-Hassan Gabarin, professor and scholar in the history of Arabic and Islamic science and technology, painstakingly researched primary Arabic source-texts revealing that each of the prerequisite technologies for *solvent vehicle alchemy* derived from Islamic alchemy and pharmacy originating circa 8th century CE and onward. Due to space constraints, this topic must remain limited to a brief gloss as it pertains to Valentine's *XII Keys*, but is expanded on in some detail in *Téchni̱. – The Art of Confecting the Philosophers Stone*, by the same author.

Overwhelming evidence suggests that Valentine had mastered *solvent vehicle alchemy*, and the use of ether and other menstrua in *The Elucidation of the XII Keys*, his *Phalaja* work and acetate tincture products for which he is most known confirm this. However, he apparently used wet chemistry innovations in *The XII Keys* only in the context of gold technology. The overall process and the use of antimony however reveal that he indeed was equally familiar with both the archetypal Alexandrian and Islamic-derived *solvent vehicle* styles of alchemical expression. The fact that he was able to shift between the two styles so fluidly reveals a depth and mastery of alchemy rarely seen in other operators. In Valentine's case, one style does not necessarily disqualify the other, and it appears he found harmony between both. This may explain why his *XII Keys* seems to adhere closely to the archetypal Chrysopœian blueprint, whereas his *Phalaja* and antimony work is such a perfect example of Islamic-derived *solvent vehicle al/chemistry*.

One aspect of confecting *The Great Stone of the Ancient Philosophers* deliberately left out of the main text is Valentine's discussion of iron or copper as suitable substitutes for the traditional gold content. His is a valid and insightful observation because both iron and copper can be converted to chloride salts (purple-red iron trichloride; $FeCl_3$ – or – anhydrous brown cupric chloride; $CuCl_2$) in the same manner as gold to yield a red colored Philosophers' Stone. It is likely that some of the more spectacular and dynamic color-changes described by European alchemists require the addition of either iron, copper or as Valentine says – "cut of both". The possibility of replacing gold with iron, copper or a combination thereof brings reproducibility experiments within the financial budget range of most experimenters and would-be adepts.

> *You will finde, that the nature of the golden sulphur consisteth only in all Metals, which are comprehended among the red, and have a fellow dominion with other Minerals, by reason of the fiery tinging spirits, but the magnetick power and its quality resteth in its white Mercurial spirit, which bindeth the Soul, and dissolveth the body, therefore the Astrum of Sol is found not only in Gold, that with the addition of the spirit of Mercury and the Solar Salt only the Philosophers stone could be made, but may in like manner be prepared artificially out of Copper and Steel [iron], two immature Metals, both which as male [♂] and female [♀] have red tinging qualities, as well as Gold itself, whither the same be taken out of one alone, or cut of both, being first ent[e]red into an Union.*

www.ingramcontent.com/pod-product-compliance
Lightning Source LLC
Chambersburg PA
CBHW050540300426
44113CB00012B/2193